INTRODUCTION TO MODERN STATISTICAL MECHANICS

INTRODUCTION TO MODERN STATISTICAL MECHANICS

David Chandler

University of California, Berkeley

New York   Oxford

OXFORD UNIVERSITY PRESS

1987

Oxford University Press

Oxford   New York   Toronto
Delhi   Bombay   Calcutta   Madras   Karachi
Petaling Jaya   Singapore   Hong Kong   Tokyo
Nairobi   Dar es Salaam   Cape Town
Melbourne   Auckland

and associated companies in
Beirut   Berlin   Ibadan   Nicosia

Published by Oxford University Press, Inc.,
200 Madison Avenue, New York, New York 10016

Oxford is a registered trademark of Oxford University Press

Library of Congress Cataloging-in-Publication Data
Chandler, David, 1944–
Introduction to modern statistical mechanics.
Includes bibliographies and index.
1. Statistical mechanics.
2. Statistical thermodynamics.
3. Chemistry, Physical and theoretical.
I. Title.
QC174.8.C4/ 1987 530.1'3 86-17950
ISBN 0-19-504276-X
ISBN 0-19-504277-8 (pbk.)

46810975
Printed in the United States of America
on acid-free paper

To the memory of W. H. Flygare

Preface

This book presents the material I use in my lectures on elementary statistical mechanics, a one-semester course I have taught at the University of Illinois and at the University of Pennsylvania. Students enter this course with some knowledge of thermodynamics, the Boltzmann distribution law, and simple quantum mechanics, usually acquired from undergraduate physical chemistry or modern physics at the level of Moore's *Physical Chemistry*. The purpose of my lectures is not only to give the students a deeper understanding of thermodynamics and the principles of equilibrium statistical mechanics, but also to introduce them to the modern topics of Monte Carlo sampling, the renormalization group theory, and the fluctuation-dissipation theorem. The ideas surrounding these topics have revolutionized the subject of statistical mechanics, and it is largely due to them that the practitioners of statistical mechanics now play a significant role in the current research and discoveries of fields ranging from molecular biology to material science and engineering, to chemical structure and dynamics, and even to high energy particle physics. Thus, in my opinion, no serious student of the natural sciences is properly educated without some understanding of these modern topics and concepts such as "order parameters" and "correlation functions." It is also my opinion and experience that these topics and concepts can and should be covered in a one-semester introductory course.

To cover this material at an introductory level, I make frequent use of simplified models. In this way, I can keep the mathematics relatively simple yet still describe many of the sophisticated ideas in the field. I refrain from treating advanced theoretical techniques (e.g., diagrammatic and field theoretic methods) even though these are among my favorite topics for research. I also treat only briefly the traditional statistical thermodynamics subjects of ideal gases and gas phase chemical equilibria. In the former case, this book should provide necessary background for the interested student to pursue advanced courses or readings in many-body theory. In the latter case, since there are already so many excellent texts devoted to these topics, it is wastefully redundant to spend much time on them here. Furthermore, these subjects have now become rather standard in the materials presented in undergraduate physical chemistry courses.

I have adopted a particular sequence of topics that perhaps deserves some comment. The first two chapters are devoted entirely to macroscopic thermodynamics, and microscopic statistical principles do not appear until Chapter 3. In Chapter 1, I review elementary thermodynamics and introduce Legendre transforms, and in Chapter 2, I develop the concepts of phase equilibria and stability. I use this strategy because the techniques and language introduced here greatly add to the swiftness with which the principles of statistical mechanics can be understood and applied. Another approach could begin with the first half of Chapter 3 where the second law appears as the direct consequence of the statistical assumption that macroscopic equilibrium is the state of greatest randomness. Then the thermodynamics in Chapters 1 and 2 could be combined with the associated material in the latter half of Chapter 3 and in Chapters 4 and 5. The different ensembles and the role of fluctuations are treated in Chapter 3. Chapters 4 and 5 refer to my discussions of the statistical mechanics of non-interacting ideal systems and phase transformations, respectively.

The treatment of phase transitions in Chapter 5 focuses on the Ising model. Within the context of that model, I discuss both mean field approximations and the renormalization group theory. In the latter case, I know of no other introductory text presenting a self-contained picture of this important subject. Yet, as I learned from Humphrey Maris and Leo Kadanoff's pedagogical article,* it is possible to teach this material at an elementary level and bring the students to the point where they can perform renormalization group calculations for problem set exercises.

Chapter 6 presents another very important subject not treated in other texts of this level—the Monte Carlo method. Here, I again use the Ising model as a concrete basis for discussion. The two-dimensional case illustrates the behavior of fluctuations in a system

---

* H. J. Maris and L. J. Kadanoff, *Am. J. Phys.* **46,** 652 (1978).

which, if large enough, could exhibit true phase equilibria and interfacial phenomena. The one-dimensional case serves to illustrate principles of quantum Monte Carlo. The Metropolis algorithm is described, and programs are provided for the student to experiment with a microcomputer and explore the power and limitations of the method.

In Chapter 7, we consider the equilibrium statistical mechanics of classical fluids. In chemistry, this is a very important topic since it provides the basis for understanding solvation. Some of the topics, such as the Maxwell-Boltzmann velocity distribution, are rather standard. But others are less so. In particular, definitions and descriptions of pair correlation functions for both simple and molecular fluids are presented, the connection between these functions and X-ray scattering cross sections is derived, and their relationship to chemical equilibria in solutions is discussed. Finally, an illustration of Monte Carlo for a two-dimensional classical fluid of hard disks is presented which the student can run on a microcomputer.

The last chapter concerns dynamics—relaxation and molecular motion in macroscopic systems that are close to or at equilibrium. In particular, I discuss time correlation functions, the fluctuation-dissipation theorem and its consequences for understanding chemical kinetics, self-diffusion, absorption, and friction. Once again, in the context of modern science, these are very important and basic topics. But in terms of teaching the principles of non-equilibrium statistical mechanics, the subject has too often been considered as an advanced or special topic. I am not sure why this is the case. A glance at Chapter 8 shows that one may derive the principal results such as the fluctuation-dissipation theorem with only a few lines of algebra, and without recourse to sophisticated mathematical methods (e.g., propagators, projectors, and complex variables).

In all the chapters, I assume the reader has mastered the mathematical methods of a typical three-semester undergraduate calculus course. With that training, the student may find some of the mathematics challenging yet manageable. In this context, the most difficult portion of the book is Chapters 3 and 4 where the concepts of probability statistics are first encountered. But since the material in these chapters is rather standard, even students with a weak background but access to a library have been able to rise to the occasion. Students who have taken the course based on this text have been advanced undergraduates or beginning graduates majoring in biochemistry, chemistry, chemical engineering, or physics. They usually master the material well enough to answer most of the numerous Exercise questions. These Exercises form an integral part of the book, reinforcing and testing every subject, and in some cases my only coverage of certain topics is found in the Exercises.

After their study of this book, I do hope a significant number of students will pursue more advanced treatments of the subjects than those I present here. For that reason, I have purposely peppered the text with comments and questions to spur the students' curiosity and perhaps send them to the library for more reading. The Bibliography at the end of each chapter suggests places for the students to start this reading. In this sense, this book serves as both an introduction and a guide to a discipline too broad and powerful for any one text to adequately describe.

In creating this book, I have benefited from the help of many people. John Wheeler has given his time unstintingly to help weed out logical errors and points of confusion. Encouragement and advice from Attila Szabo are greatly appreciated. I am also grateful to John Light for his helpful review of an earlier version of the text. Several students and my wife, Elaine, wrote and tested the computer programs included in the book. Elaine provided a great deal of advice on the content of the book as well. Finally, I am indebted to Evelyn Carlier and Sandy Smith, respectively, for their expert manuscript preparations of the first and final versions of the text.

Philadelphia and Berkeley   D. C.
January 1986

Contents

A Note to the Student

There are many Exercises to be done both within the main body of the text and at the end of each chapter. For the most part, those placed within the text are meant to be worked out immediately as simple exercises. Occasionally, however, I have engaged in the pedagogical device of asking questions that require development of a sophisticated idea in order to obtain the answer. These Exercises are marked with an asterisk, and there are three ways of handling them. First, you might figure out the answer on the spot (in which case you are more than pretty good!). Second, you could "cheat" and look for the required techniques in other textbooks (a kind of "cheating" I hope to inspire). Third, you can keep on thinking about the problem but proceed with studying the text. In the last case, you will often find the techniques for the solution will gradually appear later.

INTRODUCTION TO MODERN STATISTICAL MECHANICS

CHAPTER 1

Thermodynamics, Fundamentals

Statistical mechanics is the theory with which we analyze the behavior of natural or spontaneous fluctuations. It is the ubiquitous presence of fluctuations that makes observations interesting and worthwhile. Indeed, without such random processes, liquids would not boil, the sky would not scatter light, indeed every dynamic process in life would cease. It is also true that it is the very nature of these fluctuations that continuously drives all things toward ever-increasing chaos and the eventual demise of any structure. (Fortunately, the time scales for these eventualities are often very long, and the destruction of the world around us by natural fluctuations is not something worth worrying about.) Statistical mechanics and its macroscopic counterpart, thermodynamics, form the mathematical theory with which we can understand the magnitudes and time scales of these fluctuations, and the concomitant stability or instability of structures that spontaneous fluctuations inevitably destroy.

The presence of fluctuations is a consequence of the complexity of the systems we observe. Macroscopic systems are composed of many particles—so many particles that it is impossible to completely control or specify the system to an extent that would perfectly prescribe the evolution of the system in a deterministic fashion. Ignorance, therefore, is a law of nature for many particle systems, and this ignorance leads us to a statistical description of observations and the admittance of ever-present fluctuations.

Even those observed macroscopic properties we conceive of as being static are irrevocably tied to the statistical laws governing dynamical fluctuations. As an illustration, consider a dilute gas that obeys the ideal gas equation of state: $pV = nRT$ (p is pressure, V is the volume of the container, n is the number of moles, T is temperature, and R is the gas constant). In Chapter 3, we will show that this equation is equivalent to a formula for the mean square density fluctuations in the gas. The equation can be regarded entirely as a consequence of a particular class of statistics (in this case, the absence of correlations between density fluctuations occurring at different points in space), and need not be associated with any details of the molecular species in the system. Further, if these (uncorrelated) density fluctuations ceased to exist, the pressure would also vanish.

As we will see later in Chapter 8, we can also consider the correlation or influence of a fluctuation occurring at one instant with those occurring at other points in time, and these considerations will tell us about the process of relaxation or equilibration from nonequilibrium or unstable states of materials. But before we venture deeply into this subject of characterizing fluctuations, it is useful to begin by considering what is meant by "equilibrium" and the energetics associated with removing macroscopic systems from equilibrium. This is the subject of thermodynamics. While many readers of this book may be somewhat familiar with this subject, we take this point as our beginning because of its central importance to statistical mechanics. As we will discuss in Chapter 3, the reversible work or energetics associated with spontaneous fluctuations determines the likelihood of their occurrence. In fact, the celebrated second law of thermodynamics can be phrased as the statement that at equilibrium, all fluctuations consistent with the same energetics are equally likely. Before discussing the second law, however, we need to review the first law and some definitions too.

1.1 First Law of Thermodynamics and Equilibrium

The first law is concerned with the *internal energy*. The quantity, to which we give the symbol E, is defined as the total energy of the system. We postulate that it obeys two properties. First, internal energy is *extensive*. That means it is additive. For example, consider the composite system pictured in Fig. 1.1. By saying that the internal energy is extensive, we mean

$$E = E_1 + E_2.$$

Due to this additivity, extensive properties depend linearly on the size of the system. In other words, if we double the size of the system keeping other things fixed, the energy of the system will double.

The second postulated property is that we assume energy is

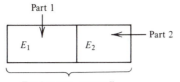

Fig. 1.1. Composite system.

conserved. This means that if the energy of a system changes, it must be as a result of doing something to the system—that is, allowing some form of energy to flow into or out of the system. One thing we can do is perform mechanical work. What else is there? Empirically we know that the energy of a system can be changed by doing work on the system, or by allowing heat to flow into the system. Therefore, we write as a definition of *heat*:

$$dE = đQ + đW.$$

This equation is usually referred to as the first law. In it, $đW$ is the differential work done *on* the system (manipulating mechanical constraints), and $đQ$ is the differential heat flow into the system. The work term has the general form

$$đW = \mathbf{f} \cdot d\mathbf{X},$$

where $\mathbf{f}$ is the applied "force," and $\mathbf{X}$ stands for a mechanical extensive variable. A familiar example is

$$đW = -p_{\text{ext}} \, dV,$$

where V is the volume of a bulk system, and p_{ext} is the external pressure. As another example,

$$đW = f \, dL,$$

where here f is the tension applied to a rubber band and L is the length of that rubber band. In general, there are many mechanical extensive variables, and their changes involve work. The abbreviated vector notation, $\mathbf{f} \cdot d\mathbf{X}$, is used to indicate all the associated work terms, $f_1 \, dX_1 + f_2 \, dX_2 + \cdots$.

The definition of heat is really not complete, however, unless we postulate a means to control it. *Adiabatic walls* are the constraints that prohibit the passage of heat into the system. Provided that one state, say A, of the system can be reached from another, call it B, by some mechanical process while the system is enclosed by an adiabatic wall, it is then possible to measure the energy difference, $E_A - E_B$, by determining the work required to pass between these states by an adiabatic process.

In this remark on the measurability of energy, we are assuming that there is an experimental means for characterizing the "state" of a system.

Another important point to keep in mind is that work and heat are forms of energy *transfer*. Once energy is transferred ($đW$ or $đQ$) it is indistinguishable from energy that might have been transferred differently. Although $đW + đQ = dE$, and there is a quantity E, there are no quantities W and Q. Hence $đW$ and $đQ$ are *inexact differentials*, and the strokes in $đW$ and $đQ$ are used to indicate this fact.

Exercise 1.1 List a few familiar examples of the two types of energy flow (e.g., two ways to melt ice—stirring or sitting in the sun).

Experimentally we know that isolated systems tend to evolve spontaneously toward simple terminal states. These states are called *equilibrium states*. By "simple" we mean that *macroscopically* they can be characterized by a small number of variables. In particular, the equilibrium state of a system is completely characterized macroscopically by specifying E and $\mathbf{X}$. For a system in which the relevant mechanical extensive variables are the volume and the numbers of molecules, the variables that characterize the system are

$$E, V, n_1, \ldots, n_j, \ldots, n_r \lessgtr r \text{ components.}$$

$\uparrow$ volume

$\uparrow$ number of moles of species j

If an electric field is applied, the total dipole of the system must be added to the list of relevant variables. (By the way, in the case of electric and magnetic fields, care is required in the development of an *extensive* electrical and magnetic energy. Can you think of the source of the difficulty? [*Hint*: Think about the spatial range of interactions between dipoles.])

Incidentally, in a completely deductive development of macroscopic thermodynamics, one should distinguish between the composition variables $n_1, n_2, \ldots, n_r$ and the mechanical extensive variables such as the volume, V. We choose to ignore the distinction in this text because one can, by use of semipermeable membranes, electrochemical cells, or phase equilibria, design experiments (real or thought) in which transfer of moles and mixing occurs with the

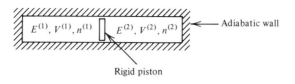

Fig. 1.2. An illustrative system.

consumption or production of work. This observation can be used to verify that composition variables play a mathematically equivalent role to the standard mechanical extensive variables. See Exercise 1.5 below.

The complete list of relevant variables is sometimes a difficult experimental issue. But whatever the list is, the most important feature of the macroscopic equilibrium state is that it is characterized by a very small number of variables, small compared to the overwhelming number of mechanical degrees of freedom that are necessary to describe in general an arbitrary non-equilibrium state of a macroscopic many particle system.

Virtually no system of physical interest is rigorously in equilibrium. However, many are in a metastable equilibrium that usually can be treated with equilibrium thermodynamics. Generally, if in the course of observing the system, it appears that the system is independent of time, independent of history, and there are no flows of energy or matter, then the system can be treated as one which is at equilibrium, and the properties of the system can be characterized by $E, V, n_1, \ldots, n_r$ alone. Ultimately, however, one is never sure that the equilibrium characterization is truly correct, and one relies on the internal consistency of equilibrium thermodynamics as a guide to the correctness of this description. An internal inconsistency is the signature of *non*-equilibrium behavior or the need for additional macroscopic variables and not a failure of thermodynamics.

What can thermodynamics tell us about these equilibrium states? Consider a system in equilibrium state I formed by placing certain constraints on the system. One (or more) of these constraints can be removed or changed and the system will evolve to a new terminal state II. The determination of state II can be viewed as the basic task of thermodynamics.

As examples, consider the system pictured in Fig. 1.2, and imagine the following possible changes:

1. Let piston move around.

2. Punch holes in piston (perhaps permeable only to one species).

3. Remove adiabatic wall and let system exchange heat with surroundings.

What terminal states will be produced as a result of these changes? To answer this question, a principle is needed. This principle is the second law of thermodynamics.

While this motivation to consider the second law is entirely macroscopic, the principle has a direct bearing on microscopic issues, or more precisely, the nature of fluctuations. The reasoning is as follows: Imagine that the constraints used to form the initial state I have just been removed and the system has begun its relaxation to state II. After the removal of the constraints, it becomes impossible to discern with certainty whether the formation of state I was the result of applied constraints (now removed) or the result of a spontaneous fluctuation. Therefore, the analysis of the basic task described above will tell us about the energetics or thermodynamics of spontaneous fluctuations, and we shall see that this information will tell us about the likelihood of fluctuations and the stability of state II.

With this foreshadowing complete, let us now turn to the principle that provides the algorithm for this analysis.

1.2 Second Law

As our remarks have already indicated, the second law is intimately related to and indeed a direct consequence of reasonable and simple statistical assumptions concerning the nature of equilibrium states. We will consider this point of view in Chapter 3. But for now, we present this law as the following postulate:

> There is an extensive function of state, $S(E, \mathbf{X})$, which is a monotonically increasing function of E, and if state B is adiabatically accessible from state A, then $S_B \geqslant S_A$.

Notice that if this state B was *reversibly* accessible from state A, then the process $B \rightarrow A$ could be carried out adiabatically too. In that case, the postulate also implies that $S_A \geqslant S_B$. Hence, if two states, A and B, are adiabatically and reversibly accessible, $S_A = S_B$. In other words, the change $\Delta S = S_B - S_A$ is zero for a reversible adiabatic process, and otherwise ΔS is positive for any natural irreversible adiabatic process. That is,

$$(\Delta S)_{\text{adiabatic}} \geqslant 0,$$

where the equality holds for reversible changes only.

The words "reversible" and "irreversible" deserve some comment. A reversible process is one that can be exactly retraced by infinitesimal changes in control variables. As such, it is a quasi-static thermodynamic process carried out arbitrarily slowly enough so that at each stage the system is in equilibrium. In other words, a reversible process progresses within the manifold of equilibrium states. Since these states are so simply characterized by a few variables, any such process can be reversed by controlling those variables; hence the name "reversible." Natural processes, on the other hand, move out of that manifold through vastly more complex non-equilibrium states requiring, in general, the listing of an enormous number of variables (perhaps the positions of all particles) to characterize these states. Without controlling all these variables (which would be impossible in a general case), it is highly unlikely that in an attempt to reverse such a process, one would observe the system passing through the same points in state space. Hence, the process is "irreversible."

The extensive function of state, $S(E, \mathbf{X})$, is called the *entropy*. As we have already demonstrated, the entropy change for a reversible adiabatic process is zero. Note also that entropy is a *function of state*. That means it is defined for those states characterized by E and $\mathbf{X}$. Such states are the thermodynamic equilibrium states. Entropy obeys several other important properties as well. To derive them, consider its differential

$$dS = (\partial S/\partial E)_\mathbf{X}\, dE + (\partial S/\partial \mathbf{X})_E \cdot d\mathbf{X},$$

where the second term is an abbreviation for $(\partial S/\partial X_1)\, dX_1 + (\partial S/\partial X_2)\, dX_2 + \cdots$. For a reversible process, we also have

$$dE = (\dbar Q)_{\text{rev}} + \mathbf{f} \cdot d\mathbf{X}.$$

Here, due to reversibility, the "force," $\mathbf{f}$, is a property of the system. For instance, at equilibrium, the externally applied pressures, p_{ext}, are the same as the pressure of the system, p.

Combining the last two equations gives

$$dS = (\partial S/\partial E)_\mathbf{X}\, (\dbar Q)_{\text{rev}} + [(\partial S/\partial \mathbf{X})_E + (\partial S/\partial E)_\mathbf{X}\mathbf{f}] \cdot d\mathbf{X}.$$

For an adiabatic process that is reversible, we have that both dS and $(\dbar Q)_{\text{rev}}$ are zero. Since the last equation must hold for all reversible processes (i.e., all displacements connecting the manifold of equilibrium states), it must hold for reversible adiabatic processes. To ensure this behavior, the term in square brackets in the last equation must be identically zero. Hence

$$(\partial S/\partial \mathbf{X})_E = -(\partial S/\partial E)_\mathbf{X}\mathbf{f}.$$

Note that all the quantities involved in this equation are functions of state. Therefore, the equality holds for nonadiabatic as well as adiabatic processes.

We postulate that S is a monotonically increasing function of E; that is, $(\partial S/\partial E)_X > 0$, or $(\partial E/\partial S)_X \geq 0$. This last derivative is defined as the temperature, T. That is,

$$T \equiv (\partial E/\partial S)_X \geq 0.$$

We will see later that this definition is consistent with our physical notion of temperature. Note that since both E and S are extensive, the temperature is *intensive*. That is, T is independent of the size of the system.

The last two equations combine to give

$$(\partial S/\partial X)_E = -\mathbf{f}/T.$$

Therefore, since $dS = (\partial S/\partial E)_X\, dE + (\partial S/\partial X)_E \cdot d\mathbf{X}$, we have

$$\boxed{dS = (1/T)\, dE - (\mathbf{f}/T) \cdot d\mathbf{X}}$$

or

$$\boxed{dE = T\, dS + \mathbf{f} \cdot d\mathbf{X}.}$$

According to this last equation, the energy for equilibrium states is characterized by S and $\mathbf{X}$, that is,

$$E = E(S, \mathbf{X}).$$

The boxed equations of this section are the fundamental relationships that constitute the mathematical statement of the second law of thermodynamics.

Exercise 1.2 An equation of state for a rubber band is either

$$S = L_0 \gamma (\theta E/L_0)^{1/2} - L_0 \gamma \left[\frac{1}{2} \left(\frac{L}{L_0} \right)^2 + \frac{L_0}{L} - \frac{3}{2} \right], \qquad L_0 = n l_0,$$

or

$$S = L_0 \gamma e^{\theta n E/L_0} - L_0 \gamma \left[\frac{1}{2} \left(\frac{L}{L_0} \right)^2 + \frac{L_0}{L} - \frac{3}{2} \right], \qquad L_0 = n l_0,$$

where γ, l_0, and θ are constants, L is the length of the rubber band, and the other symbols have their usual meaning. Which of the two possibilities is acceptable? Why? For the acceptable choice deduce the dependence of the tension, f, upon T and L/n; that is, determine $f(T, L/n)$.

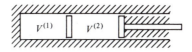

Fig. 1.3. Composite system illustrating the meaning of internal constraints.

1.3 Variational Statement of Second Law

A useful form of the second law, and one that is most closely tied to the energetics of fluctuations, is derived by considering a process in which an *internal constraint* is applied quasi-statically at constant E and $\mathbf{X}$.

Internal constraints are constraints that couple to extensive variables but do not alter the total value of those extensive variables. For example, consider the system pictured in Fig. 1.3. The total value $V = V^{(1)} + V^{(2)}$ can be changed by moving the piston on the right. But that process would *not* correspond to the application of an internal constraint. Rather, imagine moving the interior piston. It would require work to do this and the energy of the system would therefore change. But the total volume would not change. This second process does correspond to an application of an internal constraint.

With this definition in mind, consider the class of processes depicted in Fig. 1.4. The system initially is at equilibrium with entropy $S = S(E, \mathbf{X})$. Then, by applying an internal constraint, the system is reversibly brought to a constrained equilibrium with the same E and $\mathbf{X}$, but with entropy $S' = S(E, \mathbf{X};$ internal constraint). The states on the E-$\mathbf{X}$ plane are the manifold of equilibrium states in the absence of the internal constraint. The application of the internal

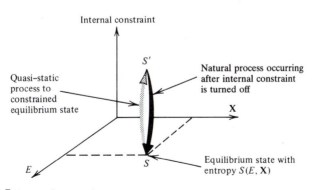

Fig. 1.4. Entropy changes for a process involving the manipulation of an internal constraint.

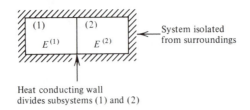

Fig. 1.5. Illustrative composite system.

constraint lifts the system off this manifold. It will require work to do this, and the requirement that there is no change in energy for the process means that there must also have been a flow of heat.

After attaining and while maintaining this constrained state, we will adiabatically insulate the system. Then, let us imagine what will happen when we suddenly shut off the internal constraint. The system will relax naturally at constant E and $\mathbf{X}$ back to the initial state with entropy S as depicted in Fig. 1.4. By considering just this last leg of this cyclic process, the second law tells us that the entropy change is positive—that is,

$$S - S' > 0,$$

or that

$$S(E, \mathbf{X}) > S(E, \mathbf{X}; \text{ internal constraint}).$$

In other words, the equilibrium state is the state at which $S(E, \mathbf{X}; \text{internal constraint})$ has its global maximum.

This variational principle provides a powerful method for deducing many of the consequences of the second law of thermodynamics. It clearly provides the algorithm by which one may solve the problem we posed as the principal task of thermodynamics. To see why, consider the example in Fig. 1.5. We can ask: Given that the system was initially prepared with E partitioned with $E^{(1)}_{\text{initial}}$ in subsystem 1 and $E^{(2)}_{\text{initial}}$ in subsystem 2, what is the final partitioning of energy? That is, when the system equilibrates, what are the values of $E^{(1)}$ and $E^{(2)}$? The answer is: $E^{(1)}$ and $E^{(2)}$ are the values that maximize $S(E, \mathbf{X}; E^{(1)}, E^{(2)})$ subject to the constraint $E^{(1)} + E^{(2)} = E$.

The entropy maximum principle has a corollary that is an *energy minimum principle*. To derive this corollary, we consider the composite system drawn above and let $E^{(1)}$ and $E^{(2)}$ denote the equilibrium partitioning of the energy. The entropy maximum principle implies

$$S(E^{(1)} - \Delta E, \mathbf{X}^{(1)}) + S(E^{(2)} + \Delta E, \mathbf{X}^{(2)}) < S(E^{(1)} + E^{(2)}, \mathbf{X}^{(1)} + \mathbf{X}^{(2)}).$$

Here, the quantity ΔE is an amount of energy removed from subsystem 1 and placed into subsystem 2. Such repartitioning lowers the entropy as indicated in the inequality. Note that in computing the entropy of the repartitioned system, we have used the fact that entropy is extensive so that we simply add the entropies of the two separate subsystems.

Now recall that S is a monotonically increasing function of E (i.e., the temperature is positive). Therefore, there is an energy

$$E < E^{(1)} + E^{(2)}$$

when $\Delta E \neq 0$ such that

$$S(E^{(1)} - \Delta E, \mathbf{X}^{(1)}) + S(E^{(2)} + \Delta E, \mathbf{X}^{(2)}) = S(E, \mathbf{X}^{(1)} + \mathbf{X}^{(2)}).$$

In other words, we can imagine applying internal constraints at constant total S and $\mathbf{X}$, and such processes will necessarily raise the total energy of the system. That is, $E(S, \mathbf{X})$ *is a global minimum of* $E(S, \mathbf{X}; internal\ constraint)$. This statement is the energy minimum principle to which we have referred.

Often, the extremum principles are stated in terms of mathematical variations away from the equilibrium state. We can write ΔE for such variations in terms of a Taylor series:

$$\Delta E = E(S, \mathbf{X}; \delta Y) - E(S, \mathbf{X}; 0)$$
$$= (\delta E)_{S,\mathbf{X}} + (\delta^2 E)_{S,\mathbf{X}} + \cdots,$$

where δY denotes a variation or partitioning of internal extensive variables caused by the application of an internal constraint, and

$$(\delta E)_{S,\mathbf{X}} = \text{first-order variational displacement}$$
$$= [(\partial E/\partial Y)_{S,\mathbf{X}}]_{Y=0}\, \delta Y$$
$$(\delta^2 E)_{S,\mathbf{X}} = \text{second-order variational displacement}$$
$$= [(1/2)(\partial^2 E/\partial Y^2)_{S,\mathbf{X}}]_{Y=0}(\delta Y)^2, \text{ etc.}$$

The principles cited above are (with this notation)

$$(\delta E)_{S,\mathbf{X}} \geq 0$$

for any small variation away from the equilibrium manifold of states with $\delta Y = 0$, and $(\Delta E)_{S,\mathbf{X}} > 0$ for all small variations away from the *stable* equilibrium state. Similarly, $(\Delta S)_{E,\mathbf{X}} < 0$.

1.4 Application: Thermal Equilibrium and Temperature

An instructive use of the variational form of the second law establishes the criterion for thermal equilibrium and identifies the integrating factor, T, as the property we really conceive of as the temperature.

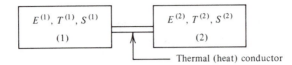

Fig. 1.6. Heat conducting system.

Consider the system in Fig. 1.6. Let us ask the question: At equilibrium, how are $T^{(1)}$ and $T^{(2)}$ related? To obtain the answer, imagine a small displacement about equilibrium due to an internal constraint. The variational theorem in the entropy representation is

$$(\delta S)_{E,\mathbf{x}} \leq 0.$$

Since $E = E^{(1)} + E^{(2)}$ is a constant during the displacement,

$$\delta E^{(1)} = -\delta E^{(2)}.$$

Further, since S is extensive

$$S = S^{(1)} + S^{(2)}.$$

Thus

$$\delta S = \delta S^{(1)} + \delta S^{(2)}$$
$$= \left(\frac{\partial S^{(1)}}{\partial E^{(1)}}\right)_{\mathbf{x}} \delta E^{(1)} + \left(\frac{\partial S^{(2)}}{\partial E^{(2)}}\right)_{\mathbf{x}} \delta E^{(2)}$$
$$= \left(\frac{1}{T^{(1)}} - \frac{1}{T^{(2)}}\right) \delta E^{(1)},$$

where in the last equality we noted that

$$(\partial S / \partial E)_{\mathbf{x}} = 1/T$$

and applied the condition that $\delta E^{(2)} = -\delta E^{(1)}$. Hence,

$$\left(\frac{1}{T^{(1)}} - \frac{1}{T^{(2)}}\right) \delta E^{(1)} \leq 0$$

for *all* (positive or negative) small variations $\delta E^{(1)}$. The *only* way this condition can hold is if

$$\boxed{T^{(1)} = T^{(2)}}$$

at equilibrium.

Notice the perspective we have adopted to derive the equilibrium condition of equal temperatures: We assume the system is at an equilibrium state, and we learn about that state by perturbing the system with the application of an internal constraint. We then analyze the response to that perturbation, and by comparing with the second law arrive at a conclusion concerning the equilibrium state we began with. This type of procedure is very powerful and we will use it often. In a real sense, it is closely tied to the nature of experimental observations. In particular, if we conceive of a real experiment that will probe the behavior of an equilibrium system, the experiment will generally consist of observing the response of the system to an imposed perturbation.

With this procedure, we have just proven that according to the second law, the criterion for thermal equilibrium is that the two interacting subsystems have the same temperature. Our proof used the entropy variational principle. You can verify that the same result follows from the energy variational principle.

Exercise 1.3 Perform the verification.

Next let's change our perspective and consider what happens when the system is initially not at thermal equilibrium—that is, initially $T^{(1)} \neq T^{(2)}$. Eventually, at thermal equilibrium, the two become equal. The question is: how does this equilibrium occur?

To answer the question we use the second law again, this time noting that the passage to equilibrium is a natural process so that the change in entropy, ΔS, is positive (i.e., $dS > T^{-1} \, \dbar Q$ and $\dbar Q$ for the total system is zero). Thus

$$\Delta S^{(1)} + \Delta S^{(2)} = \Delta S > 0.$$

As a result (assuming differences are small)

$$\left(\frac{\partial S^{(1)}}{\partial E^{(1)}} \right)_{\mathbf{x}} \Delta E^{(1)} + \left(\frac{\partial S^{(2)}}{\partial E^{(2)}} \right)_{\mathbf{x}} \Delta E^{(2)} > 0,$$

or noting $\Delta E^{(1)} = -\Delta E^{(2)}$ and $(\partial S/\partial E)_{\mathbf{x}} = 1/T$,

$$\left(\frac{1}{T^{(1)}} - \frac{1}{T^{(2)}} \right) \Delta E^{(1)} > 0.$$

Now suppose $T^{(1)} > T^{(2)}$. Then $\Delta E^{(1)} < 0$ to satisfy the inequality. Similarly if $T^{(1)} < T^{(2)}$, then we must have $\Delta E^{(1)} > 0$. Thus, we have just proven that energy flow is from the hot body to the cold body.

In summary, therefore, heat is that form of energy flow due to a temperature gradient, and the flow is from hot (higher T) to cold (lower T).

Since heat flow and temperature changes are intimately related, it is useful to quantify their connection by introducing *heat capacities*.

For a quasi-static process we can define

$$C = \frac{\dd Q}{dT} = T\frac{\dd Q/T}{dT} = T\frac{dS}{dT}.$$

Of course, we should really specify the direction in which these differential changes occur. The conventional and operational definitions are

$$C_f = T\left(\frac{\partial S}{\partial T}\right)_f \quad \text{and} \quad C_\mathbf{X} = T\left(\frac{\partial S}{\partial T}\right)_\mathbf{X}.$$

Since S is extensive, C_f and $C_\mathbf{X}$ are extensive.

Exercise 1.4 Suppose you have two pieces of rubber band, each one obeying the equation of state studied in Exercise 1.2. The temperature, length per mole, and mole number for the first piece are $T^{(1)}$, $l^{(1)}$, and $n^{(1)}$, respectively. Similarly, the second piece has $T^{(2)}$, $l^{(2)}$, and $n^{(2)}$. Determine (as a function of these initial thermodynamic properties) the final energies and temperatures of the two rubber bands if they are held at constant length and placed in thermal contact with one another. Neglect thermal convection to the surroundings and mass flow.

1.5 Auxiliary Functions and Legendre Transforms

In the previous sections, we have introduced all the principles necessary for analyzing the macroscopic thermodynamics of any system. However, the ease with which the analyses can be performed is greatly facilitated by introducing certain mathematical concepts and methods. The tricks we learn here and in the remaining sections of this chapter will also significantly add to our computational dexterity when performing statistical mechanical calculations.

The first of these methods is the procedure of *Legendre transforms*. To see why the method is useful, let's use the specific form of the reversible work differential appropriate to systems usually studied in chemistry and biology. In that case, for reversible displacements

$$\mathbf{f} \cdot d\mathbf{X} = -p\,dV + \sum_{i=1}^{r} \mu_i\,dn_i,$$

where p is the pressure of the system, V is the volume of the system, n_i is the number of moles of species i (there are r such species), and μ_i is the *chemical potential* of species i. The chemical potential is defined by the equation above. It is the reversible rate of change of

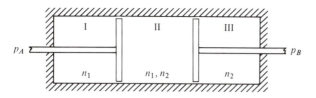

Fig. 1.7. Composite system illustrating how mechanical work can be associated with changes in mole numbers.

internal energy caused by changing mole number n_i while keeping all other mole numbers, S and V, constant. As we will stress in Chapters 2 and 3, the chemical potential is the intensive property that controls mass or particle equilibrium just as temperature controls thermal equilibrium. Indeed, we will find that gradients in chemical potentials will induce the flow of mass or the rearrangements of atoms and molecules, and the absence of such gradients ensures mass equilibrium. Rearrangements of atoms and molecules are the processes by which equilibria between different phases of matter and between different chemical species are established. Chemical potentials will therefore play a central role in many of our considerations throughout this book.

Exercise 1.5 Consider the system pictured in Fig. 1.7. The piston between subsystems I and II is permeable to species 1 but not to species 2. It is held in place with the application of pressure p_A. The other piston, held in place with pressure p_B, is permeable to species 2 but not to species 1. Note that at equilibrium, $p_A = p_{II} - p_I$ and $p_B = p_{II} - p_{III}$. Show that with appropriate constraints (e.g., surrounding the system with adiabatic walls), the reversible mechanical work performed on the system by controlling pressures p_A and p_B is reversible work we can associate with changing the concentration variables in subsystem II. Can you think of other devices for which reversible work is connected with changing mole numbers in a system?

In view of the form for $\mathbf{f} \cdot d\mathbf{X}$, we have

$$dE = T\,dS - p\,dV + \sum_{i=1}^{r} \mu_i\,dn_i.$$

Thus, $E = E(S, V, n_1, \ldots, n_r)$ is a natural function of S, V, and the n_i's. As a result, the variational principles $(\Delta E)_{S,V,n} > 0$ and $(\delta E)_{S,V,n} \geq 0$ tell us facts about equilibrium states characterized with S, V, and n_i's. But now suppose an experimentalist insists upon using T, V, and n to characterize the equilibrium states of a one-component system. The questions are: is that characterization possible, and, if so, what is the thermodynamic function analogous to E that is a natural function of T, V, and n?

The answer comes from considering Legendre transformations. We do this now. Suppose $f = f(x_1, \ldots, x_n)$ is a natural function of $x_1, \ldots, x_n$. Then

$$df = \sum_{i=1}^{n} u_i \, dx_i, \qquad u_i = (\partial f / \partial x_i)_{x_j}.$$

Let

$$g = f - \sum_{i=r+1}^{n} u_i x_i.$$

Clearly,

$$dg = df - \sum_{i=r+1}^{n} [u_i \, dx_i + x_i \, du_i]$$

$$= \sum_{i=1}^{r} u_i \, dx_i + \sum_{i=r+1}^{n} (-x_i) \, du_i.$$

Thus, $g = g(x_1, \ldots, x_r, u_{r+1}, \ldots, u_n)$ is a natural function of $x_1, \ldots, x_r$ and the conjugate variables to $x_{r+1}, \ldots, x_n$, namely, $u_{r+1}, \ldots, u_n$. The function g is called a Legendre transform of f. It transforms away the dependence upon $x_{r+1}, \ldots, x_n$ to a dependence upon $u_{r+1}, \ldots, u_n$. It is apparent that this type of construction provides a scheme for introducing a natural function of T, V, and n since T is simply the conjugate variable to S. But for the scheme to provide a satisfactory solution to the questions posed above, the Legendre transform, $g(x_1, \ldots, x_r, u_{r+1}, \ldots, u_n)$, must contain no more or less information than $f(x_1, \ldots, x_n)$. This equivalency is readily established by noting that one may always return to the original function $f(x_1, \ldots, x_n)$. Geometrically, this observation corresponds to the fact that within a constant, a function can be equivalently represented either as the envelope of a family of tangent lines or by the locus of points satisfying $f = f(x_1, \ldots, x_n)$.

To construct a natural function of T, V, and n, we thus subtract from $E(S, V, n)$ the quantity $S \times$ (variable conjugate to S) $= ST$. Hence, we let

$$A = E - TS = A(T, V, n),$$

which is called the *Helmholtz free energy*. Clearly,

$$dA = -S\,dT - p\,dV + \sum_{i=1}^{r} \mu_i\,dn_i.$$

The Legendre transform allows us to exchange a thermodynamic variable with its conjugate. Exchanges between non-conjugate pairs, however, are not possible for this scheme. For example, since (S, V, n) provides enough information to characterize an equilibrium state, so do (T, V, n), (S, p, n), and (T, p, n). However, (p, v, n) or (S, T, p) are *not* examples of sets of variables that can characterize a state. We will return to this point again when discussing thermodynamic degrees of freedom and the Gibbs phase rule.

Other possible Legendre transforms are

$$G = E - TS - (-pV)$$
$$= E - TS + pV = G(T, p, n),$$

and

$$H = E - (-pV) = E + pV$$
$$= H(S, p, n),$$

which are called the Gibbs free energy and enthalpy, respectively. Their differentials are

$$dG = -S\,dT + V\,dp + \sum_{i=1}^{r} \mu_i\,dn_i.$$

and

$$dH = T\,dS + V\,dp + \sum_{i=1}^{r} \mu_i\,dn_i.$$

Exercise 1.6 Construct Legendre transforms of the entropy that are natural functions of $(1/T, V, n)$ and of $(1/T, V, \mu/T)$.

The variational principles associated with the auxiliary functions H, A, and G are

$$(\Delta H)_{S,p,n} > 0, \qquad (\delta H)_{S,p,n} \geq 0,$$
$$(\Delta A)_{T,V,n} > 0, \qquad (\delta A)_{T,V,n} \geq 0,$$
$$(\Delta G)_{T,p,n} > 0, \qquad (\delta G)_{T,p,n} \geq 0,$$

Exercise 1.7 Derive these principles.

In Chapter 2, we will use these various representations of the second law to derive conditions that characterize stable equilibrium states.

1.6 Maxwell Relations

Armed with the auxiliary functions, many types of different measurements can be interrelated. As an example, consider

$(\partial S / \partial V)_{T,n}.$

implies we are viewing S as a function of v, T, n.

To analyze this derivative, we look at the differential of the natural function of T, V, and n:

$$dA = -S\,dT - p\,dV + \mu\,dn.$$

Note that if $df = a\,dx + b\,dy$, then $(\partial a / \partial y)_x = (\partial b / \partial x)_y$. As a result, the differential for A implies

$$(\partial S / \partial V)_{T,n} = (\partial p / \partial T)_{V,n},$$

which is a Maxwell relation.

As another example, consider $(\partial S / \partial p)_{T,n}$, which requires us to view S as a function of p, T, and n. Thus, look at the differential for G,

$$dG = -S\,dT + V\,dp + \mu\,dn.$$

Once again, noting that if $a\,dx + b\,dy$ is an exact differential, then $(\partial a / \partial y)_x = (\partial b / \partial x)_y$; as a result, the differential for G yields

$$(\partial S / \partial p)_{T,n} = -(\partial V / \partial T)_{p,n},$$

which is another Maxwell relation.

Two more examples further demonstrate these methods and serve to show how thermal experiments are intimately linked with equation of state measurements.

EXAMPLE 1: Let

$$C_v = T\left(\frac{\partial S}{\partial T}\right)_{V,n}.$$

Then

$$\left(\frac{\partial C_v}{\partial V}\right)_{T,n} = T\left(\frac{\partial}{\partial V}\left(\frac{\partial S}{\partial T}\right)_{V,n}\right)_{T,n}$$

$$= T\left(\frac{\partial}{\partial T}\left(\frac{\partial S}{\partial V}\right)_{T,n}\right)_{V,n}$$

$$= T\left(\frac{\partial}{\partial T}\left(\frac{\partial p}{\partial T}\right)_{V,n}\right)_{V,n}$$

$$= T\left(\frac{\partial^2 p}{\partial T^2}\right)_{V,n}.$$

Exercise 1.8 Derive an analogous formula for $(\partial C_p/\partial p)_{T,n}$.

EXAMPLE 2: Let

$$C_p = T\left(\frac{\partial S}{\partial T}\right)_{p,n}.$$

Viewing S as a function of T, V, and n gives

$$(dS)_n = \left(\frac{\partial S}{\partial T}\right)_{V,n}(dT)_n + \left(\frac{\partial S}{\partial V}\right)_{T,n}(dV)_n,$$

which leads to

$$\left(\frac{\partial S}{\partial T}\right)_{p,n} = \left(\frac{\partial S}{\partial T}\right)_{V,n} + \left(\frac{\partial S}{\partial V}\right)_{T,n}\left(\frac{\partial V}{\partial T}\right)_{n,p}.$$

Hence

$$\frac{1}{T}C_p = \frac{1}{T}C_v + \left(\frac{\partial p}{\partial T}\right)_{V,n}\left(\frac{\partial V}{\partial T}\right)_{n,p}.$$

Next note that

$$\left(\frac{\partial x}{\partial y}\right)_z = -\left(\frac{\partial x}{\partial z}\right)_y\left(\frac{\partial z}{\partial y}\right)_x.$$

Exercise 1.9 Derive this formula. [*Hint:* Begin by conceiving of z as a function of x and y—that is, $dz = (\partial z/\partial x)_y\, dx + (\partial z/\partial y)_x\, dy$.]

As a result,

$$(\partial p/\partial T)_{V,n} = -(\partial p/\partial V)_{T,n}(\partial V/\partial T)_{p,n},$$

Thus

$$C_p - C_v = -T\left(\frac{\partial p}{\partial V}\right)_{T,n}\left[\left(\frac{\partial V}{\partial T}\right)_{p,n}\right]^2,$$

which is a famous result relating heat capacities to the isothermal compressibility and the coefficient of thermal expansion.

1.7 Extensive Functions and the Gibbs–Duhem Equation

We have already discussed the meaning of "extensive." In particular, a macroscopic property is extensive if it depends *linearly* on the size of the system. With this meaning in mind, consider the internal energy E, which is extensive, and how it depends upon S and $\mathbf{X}$, which are also extensive. Clearly,

$$E(\lambda S, \lambda \mathbf{X}) = \lambda E(S, \mathbf{X})$$

for any λ. Thus, $E(S, \mathbf{X})$ is a *first-order homogeneous function of S and $\mathbf{X}$.*[*] We shall derive a theorem about such functions and then use the theorem to derive the Gibbs–Duhem equation.

To begin, suppose $f(x_1, \ldots, x_n)$ is a first-order homogeneous function of $x_1, \ldots, x_n$. Let $u_i = \lambda x_i$. Then

$$f(u_1, \ldots, u_n) = \lambda f(x_1, \ldots, x_n).$$

Differentiate this equation with respect to λ:

$$\left(\frac{\partial f(u_1, \ldots, u_n)}{\partial \lambda}\right)_{x_i} = f(x_1, \ldots, x_n). \tag{a}$$

But we also know that

$$df(u_1, \ldots, u_n) = \sum_{i=1}^{n} (\partial f/\partial u_i)_{u_j} du_i,$$

and as a result

$$(\partial f/\partial \lambda)_{x_i} = \sum_{i=1}^{n} (\partial f/\partial u_i)_{u_j}(\partial u_i/\partial \lambda)_{x_i}.$$

$$= \sum_{i=1}^{n} (\partial f/\partial u_i)_{u_j} x_i. \tag{b}$$

---

[*] A homogeneous function of degree n is one for which $f(\lambda x) = \lambda^n f(x)$.

By combining Eqs. (a) and (b) we obtain

$$f(x_1, \ldots, x_n) = \sum_{i=1}^{n} (\partial f / \partial u_i)_{u_j} x_i$$

for all λ. Take $\lambda = 1$, and we have

$$f(x_1, \ldots, x_n) = \sum_{i=1}^{n} (\partial f / \partial x_i)_{x_j} x_i,$$

which is the result known as Euler's theorem for first-order homogeneous functions.

Exercise 1.10 Derive an analogous theorem for homogeneous functions of order n.

Since $E = E(S, \mathbf{X})$ is first-order homogeneous, Euler's theorem gives

$$E = (\partial E / \partial S)_{\mathbf{X}} S + (\partial E / \partial \mathbf{X})_S \cdot \mathbf{X}$$
$$= TS + \mathbf{f} \cdot \mathbf{X}.$$

Let us now specialize as in the previous two sections:

$$\mathbf{f} \cdot d\mathbf{X} = -p \, dV + \sum_{i=1}^{r} \mu_i \, dn_i.$$

Then

$$dE = T \, dS - p \, dV + \sum_{i=1}^{r} \mu_i \, dn_i, \qquad (c)$$

that is, $E = E(S, V, n_1, \ldots, n_r)$, and Euler's theorem yields

$$E = TS - pV + \sum_{i=1}^{r} \mu_i n_i.$$

Notice that the total differential of this formula is

$$dE = T \, dS + S \, dT - p \, dV - V \, dp + \sum_{i=1}^{r} [\mu_i \, dn_i + n_i \, d\mu_i]. \qquad (d)$$

By comparing Eqs. (d) and (c) we discover

$$0 = S \, dT - V \, dp + \sum_{i=1}^{r} n_i \, d\mu_i,$$

which is the *Gibbs–Duhem equation.*

Now let's apply Euler's theorem for first-order homogeneous

functions to the Gibbs free energy:

$$G = E - TS + pV$$

$$= \left(TS - pV + \sum_{i=1}^{r} \mu_i n_i \right) - TS + pV$$

$$= \sum_{i=1}^{r} \mu_i n_i.$$

Thus, for a one-component system, μ is simply the Gibbs free energy per mole, G/n.

The result $G = \sum_i \mu_i n_i$ is a special case of the general result

$$X(T, p, n_1, \ldots, n_r) = \sum_{i=1}^{r} x_i n_i,$$

where X is an extensive function when T, p, and the n_i's are used to characterize states, and the x_i's are *partial molar X's*,

$$x_i = (\partial X/\partial n_i)_{T,p,n_j} = x_i(T, p, n_1, \ldots, n_r).$$

The proof of the result that $X = \sum_i x_i n_i$ follows directly from Euler's theorem since at fixed T and p, $X(T, p, n_1, \ldots, n_r)$ is first-order homogeneous in the mole numbers:

$$X(T, p, \lambda n_1, \ldots, \lambda n_r) = \lambda X(T, p, n_1, \ldots, n_r).$$

1.8 Intensive Functions

These are zeroth-order homogeneous functions of the extensive variables. For example,

$$p = p(S, V, n_i, \ldots, n_r) = p(\lambda S, \lambda V, \lambda n_i, \ldots, \lambda n_r).$$

Exercise 1.11 Show that if X and Y are extensive, then X/Y and $\partial X/\partial Y$ are intensive.

This equation is true for any λ. Pick $\lambda^{-1} = n_1 + n_2 + \cdots + n_r = $ total number of moles $= n$. Then

$$p = p(S/n, V/n, x_1, x_2, \ldots, x_r),$$

where $x_i = n_i/n$ is the mole fraction of species i. But the different x_i's are not completely independent since

$$1 = \sum_{i=1}^{r} x_i.$$

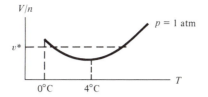

Fig. 1.8. Volume per mole of water at 1 atm pressure.

Thus,

$$p = p(S/n, V/n, x_1, \ldots, x_{r-1}, 1 - x_1 - x_2 - \cdots - x_{r-1}),$$

which proves that while $2 + r$ extensive variables are required to determine the value of an extensive property of an equilibrium system, only $1 + r$ intensive variables are needed to determine the values of other intensive variables.

The reduction from $2 + r$ to $1 + r$ degrees of freedom is simply the statement that intensive properties are independent of system size. Hence, a variable indicating this size, required for extensive properties, is not needed to characterize intensive properties.

Having said all these things, here is a simple puzzle, the solution of which provides a helpful illustration: If an intensive property of a one-component equilibrium system can be characterized by two other intensive properties, how do you explain the experimental behavior of liquid water illustrated in Fig. 1.8? The temperature T is not uniquely defined by specifying $(V/n) = v^*$ and $p = 1$ atm. The answer to this apparent paradox is that p and V are conjugate variables. Hence (p, V, n) do not uniquely characterize an equilibrium state. However, (p, T, n) do characterize a state, and so do (v, T, n). Thus, we should be able to uniquely fix v with p and T, and (v, T) should determine p. The above experimental curve does not contradict this expectation.

This puzzle illustrates the importance of using non-conjugate variables to characterize equilibrium states.

Additional Exercises

1.12. Consider a rubber band of length L held at tension f. For displacements between equilibrium states

$$dE = T\,dS + f\,dL + \mu\,dn,$$

where μ is the chemical potential of a rubber band and n is the mass or mole number of the rubber band. Derive the analog of the Gibbs–Duhem equation for a rubber band.

1.13. Suppose an equation of state for a rubber band is $E = \theta S^2 L/n^2$, where θ is a constant, L is the length of the rubber band, and the other symbols have their usual meaning. Determine the chemical potential, $\mu(T, L/n)$, and show that the equation of state satisfies the analog of the Gibbs–Duhem equation.

1.14. Show that for a one component $p - V - n$ system

$$(\partial\mu/\partial v)_T = v(\partial p/\partial v)_T,$$

where v is the volume per mole. [*Hint:* Show that $d\mu = -s\,dT + v\,dp$, where s is the entropy per mole.]

1.15. For a $p - V - n$ system, construct Legendre transforms of the entropy which are natural functions of $(1/T, V, n)$ and of $(1/T, V, \mu/T)$. Show that for a one-component system,

$$\left(\frac{\partial E}{\partial \beta}\right)_{\beta\mu, V} = -\left(\frac{\partial E}{\partial n}\right)_{\beta, V}\left(\frac{\partial n}{\partial \beta\mu}\right)_{\beta, V}\left(\frac{\partial \beta\mu}{\partial \beta}\right)_{n, V} + \left(\frac{\partial E}{\partial \beta}\right)_{n, V},$$

where $\beta = 1/T$.

1.16. Suppose a particular type of rubber band has the equation of state $l = \theta f/T$, where l is the length per unit mass, f is the tension, T is the temperature, and θ is a constant. For this type of rubber band, compute $(\partial c_l/\partial l)_T$, where c_l is the constant length heat capacity per unit mass.

1.17. Imagine holding a rubber band in a heat bath of temperature T and suddenly increasing the tension from f to $f + \Delta f$. After equilibrating, the entropy of the rubber band, $S(T, f + \Delta f, n)$, will differ from its initial value $S(T, f, n)$. Compute the change in entropy per unit mass for this process assuming the rubber band obeys the equation of state given in Exercise 1.16.

1.18. Given the generalized homogeneous function

$$f(\lambda^{\theta_1}x_1, \lambda^{\theta_2}x_2, \ldots, \lambda^{\theta_n}x_n) = \lambda f(x_1, \ldots, x_n),$$

show that

$$\theta_1 x_1\left(\frac{\partial f}{\partial x_1}\right)_{x_2, \ldots, x_n} + \cdots + \theta_n x_n\left(\frac{\partial f}{\partial x_n}\right)_{x_1, \ldots, x_{n-1}}$$
$$= f(x_1, \ldots, x_n).$$

This result is a generalization of the Euler's theorem discussed in the text. It plays an important role in the theory of phase transitions when deriving what are known as "scaling" equations of state for systems near critical points.

Bibliography

Herbert Callen pioneered the teaching of thermodynamics through a sequence of postulates, relying on the student's anticipation of a statistical microscopic basis to the principles. Our discussion in this chapter is strongly influenced by Callen's first edition:

H. B. Callen, *Thermodynamics* (John Wiley, N.Y., 1960).

In traditional treatments of thermodynamics, not a word is spoken about molecules, and the concept of temperature and the second law are motivated entirely on the basis of macroscopic observations. Here are two texts that embrace the traditional approach:

J. G. Kirkwood and I. Oppenheim, *Chemical Thermodynamics* (McGraw-Hill, N.Y., 1960),

and

J. Beatte and I. Oppenheim, *Thermodynamics* (Elsevier Scientific, N.Y., 1979).

Elementary texts are always useful. Here is a good one:

K. Denbigh, *Principles of Chemical Equilibrium, 3rd edition* (Cambridge University, Cambridge, 1971).

CHAPTER 2

Conditions for Equilibrium and Stability

In this chapter, we derive general criteria characterizing the equilibrium and stability of macroscopic systems. The derivations are based on the procedure already introduced in Chapter 1. In particular, we first assume the system under investigation is stable and at equilibrium, and we then examine the thermodynamic changes or response produced by perturbing the system away from equilibrium. The perturbations are produced by applying so-called "internal constraints." That is, the system is removed from equilibrium by repartitioning the extensive variables within the system. According to the second law of thermodynamics, such processes lead to lower entropy or higher (free) energy provided the system was initially at a stable equilibrium point. Thus, by analyzing the signs of thermodynamic changes for the processes, we arrive at inequalities consistent with stability and equilibrium. These conditions are known as equilibrium and stability criteria.

We first discuss equilibrium criteria and show, for example, that having T, p, and μ constant throughout a system is equivalent to assuring that entropy or the thermodynamic energies are extrema with respect to the partitioning of extensive variables. To distinguish between maxima or minima, one must continue the analysis to consider the sign of the curvature at the extrema. This second step in the development yields stability criteria. For example, we show that for any stable system, $(\partial T/\partial S)_{V,n} \geq 0$, $(\partial p/\partial V)_{T,n} \leq 0$, and many other similar results. In particular, stability concerns the signs of derivatives of intensive variables with respect to conjugate extensive variables, or equivalently, the signs of second derivatives of free energies with respect to extensive variables. When developing the theory of statistical mechanics in Chapter 3, we will find that these criteria, often referred to as "convexity" properties, can be viewed as statistical principles that equate thermodynamic derivatives to the value of mean square fluctuations of dynamical quantities.

After examining several thermodynamic consequences of equilibrium and stability, we apply these criteria to the phenomena of phase transitions and phase equilibria. The Clausius–Clapeyron equation and the Maxwell construction are derived. These are the thermodynamic relationships describing how, for example, the boiling temperature is determined by the latent heat and molar volume change associated with the liquid-gas phase transition. While results of this type are insightful, it should be appreciated that a full understanding of phase transitions requires a statistical mechanical treatment. Phase transformations are the results of microscopic fluctuations that, under certain circumstances, conspire in concert to bring about a macroscopic change. The explanation of this cooperativity, which is one of the great successes of statistical mechanics, is discussed in Chapter 5.

2.1 Multiphase Equilibrium

To begin the macroscopic analysis, consider a heterogeneous (multiphase) multicomponent system. Each phase comprises a different subsystem. Repartitionings of extensive variables can be accomplished by shuffling portions of the extensive variables between the different phases. For example, since E is extensive, the total energy is

$$E = \sum_{\alpha=1}^{v} E^{(\alpha)},$$

where α labels the phase and v is the total number of such phases. A repartitioning of the energy would correspond to changing the $E^{(\alpha)}$'s but keeping the total E fixed.

Here, one might note that this formula neglects energies associated with surfaces (i.e., interfaces between phases and between the system and its boundaries). The neglect of surface energy is in fact an approximation. It produces a negligible error, however, when considering large (i.e., macroscopic) bulk phases. The reason is that the energy of a bulk phase is proportional to N, the number of molecules in the phase, while the surface energy goes as $N^{2/3}$. Hence, the ratio of surface energy to bulk energy is $N^{-1/3}$, which is negligible for $N \sim 10^{24}$, the number of molecules in a mole. There are, of

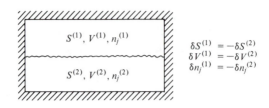

Fig. 2.1. A two-phase system.

course, situations where the surface is of interest, and a discussion of surface energy is given at the end of this chapter. But for now, we focus on the bulk phases.

Keeping only the bulk terms for entropy, too, we write

$$S = \sum_{\alpha=1}^{v} S^{(\alpha)},$$

and similarly

$$V = \sum_{\alpha=1}^{v} V^{(\alpha)}$$

and

$$n_i = \sum_{\alpha=1}^{v} n_i^{(\alpha)},$$

where $n_i^{(\alpha)}$ is the number of moles of species i in phase α. Then, from the definition of δE as the first-order variational displacement of E,

$$\delta E = \sum_{\alpha=1}^{v} \left[T^{(\alpha)} \, \delta S^{(\alpha)} - p^{(\alpha)} \, \delta V^{(\alpha)} + \sum_{i=1}^{r} \mu_i^{(\alpha)} \, \delta n_i^{(\alpha)} \right].$$

The condition for equilibrium is

$$(\delta E)_{S,V,n_i} \geq 0.$$

The subscripts indicate that we must consider processes that repartition $S^{(\alpha)}$, $V^{(\alpha)}$, and $n_i^{(\alpha)}$ keeping the total S, V, and n_i's fixed. The constancy of S, V, and the n_i's requires

$$\sum_{\alpha=1}^{v} \delta S^{(\alpha)} = 0, \qquad \sum_{\alpha=1}^{v} \delta V^{(\alpha)} = 0$$

and

$$\sum_{\alpha=1}^{v} \delta n_i^{(\alpha)} = 0 \quad \text{for} \quad i = 1, 2, \ldots, r.$$

Consider the case with $v = 2$ as pictured schematically in Fig. 2.1.

The constancy of S, V, and n_i corresponds to

$$\delta S^{(1)} = -\delta S^{(2)},$$
$$\delta V^{(1)} = -\delta V^{(2)},$$

and

$$\delta n_j^{(1)} = -\delta n_j^{(2)}.$$

The first-order displacement of E at constant S, V, and n_i is

$$0 \leqslant (\delta E)_{S,V,n_i} = (T^{(1)} - T^{(2)})\, \delta S^{(1)} - (p^{(1)} - p^{(2)})\, \delta V^{(1)}$$
$$+ \sum_{i=1}^{r} (\mu_i^{(1)} - \mu_i^{(2)})\, \delta n_i^{(1)}.$$

Note: $(\delta E)_{S,V,n_i} \geqslant 0$ must hold for *all* possible small variations $\delta S^{(1)}$, $\delta V^{(1)}$, $\delta n_i^{(1)}$. Since these variations are uncoupled and can be either positive or negative, the only acceptable solution to $(\delta E)_{S,V,n_i} \geqslant 0$ is

$$T^{(1)} = T^{(2)}, \qquad p^{(1)} = p^{(2)}$$

and

$$\mu_i^{(1)} = \mu_i^{(2)}, \qquad i = 1, 2, \ldots, r,$$

which guarantees

$$(\delta E)_{S,V,n_i} = 0$$

for small displacements away from equilibrium. Note that if the fluctuations were to be constrained to be of only one sign, then the equilibrium conditions would be inequalities rather than equalities. For example, if $\delta V^{(1)}$ could not be negative, then the analysis we have outlined would lead to the requirement that $p^{(1)} \leqslant p^{(2)}$.

The argument for unconstrained variations or fluctuations is easily extended to any number of phases. Thus, for example, if all the phases are in thermal equilibrium

$$T^{(1)} = T^{(2)} = T^{(3)} = \cdots,$$

if they are all in mechanical equilibrium

$$p^{(1)} = p^{(2)} = p^{(3)} = \cdots,$$

and if they are all in mass equilibrium

$$\mu_i^{(1)} = \mu_i^{(2)} = \mu_i^{(3)} = \cdots.$$

Exercise 2.1 Carry out the extension to verify these results.

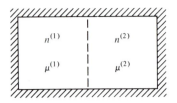

Fig. 2.2. Composite system.

It also follows that T, p, and the μ_i's are constant throughout a single homogeneous phase.

Exercise 2.2 Derive this fact.

Finally, we have shown that $(\delta E)_{S,V,n_i} \geq 0$ leads to a set of criteria for equilibrium. One may also demonstrate that these criteria are both necessary and sufficient conditions for equilibrium.

Exercise 2.3 Carry out the demonstration.

At this point, let us pause to consider a bit more about the meaning and significance of chemical potentials. Since $\mu^{(1)} = \mu^{(2)}$ guarantees mass equilibrium, it is interesting to study what action a gradient in μ produces. To do this, consider the composite system in Fig. 2.2, and suppose the system is prepared initially with $\mu^{(1)} > \mu^{(2)}$. Mass flow will bring it to equilibrium with $\mu^{(1)}_{\text{final}} = \mu^{(2)}_{\text{final}}$. If no work is done on the total system and there is no heat flow into the system,

$$\Delta S > 0$$

for the equilibration process. Assuming displacements from equilibrium are small,

$$\Delta S = -\frac{\mu^{(1)}}{T}\Delta n^{(1)} - \frac{\mu^{(2)}}{T}\Delta n^{(2)} = -\left(\frac{\mu^{(1)}}{T} - \frac{\mu^{(2)}}{T}\right)\Delta n^{(1)},$$

where $\Delta n^{(1)} = -\Delta n^{(2)}$ is the change in moles in subsystem (1) during the process. Thus, given $\mu^{(1)} > \mu^{(2)}$, $\Delta S > 0$ implies $\Delta n^{(1)} < 0$. That is, matter flows from high μ to low μ.

We see that gradients in μ (or more precisely, gradients in μ/T) produce mass flow. In that sense, $-\nabla(\mu/T)$ is a generalized force. Similarly, $-\nabla(1/T)$ is a generalized force that causes heat to flow (see Sec. 1.4). The intensive properties whose gradients cause flow of the conjugate variables are often called thermodynamic *fields* or thermodynamic *affinities*.

As a final remark concerning chemical potentials, note that we have only imagined the repartitioning of species among different phases or regions of space in the system. Another possibility is the rearrangements accomplished via chemical reactions. The analysis of these processes leads to the criteria of chemical equilibria. For now, we leave this analysis as an exercise for the reader; it will be performed in the text, however, when we treat the statistical mechanics of gas phase chemical equilibria in Chapter 4.

2.2 Stability

The condition for stable equilibrium is $(\Delta E)_{S,V,n} > 0$ for all displacements away from the equilibrium manifold of states. Hence, for small enough displacements $(\delta E)_{S,V,n} \geq 0$. But, in the previous section we discovered that for unconstrained systems (for which the internal extensive variables can fluctuate in both the positive and negative directions), $(\delta E)_{S,V,n} = 0$. Thus, at and near equilibrium

$$(\Delta E)_{S,V,n} = (\delta^2 E)_{S,V,n} + (\delta^3 E)_{S,V,n} + \cdots.$$

Since the quadratic (second-order term) will dominate for small enough displacements, we have

$$(\delta^2 E)_{S,V,n} \geq 0.$$

Conditions derived from this relation are called *stability criteria*.

If the inequality

$$(\delta^2 E)_{S,V,n} > 0$$

is satisfied, the system is stable to small fluctuations from equilibrium. That is, after the occurrence of a small fluctuation, the system will return to the equilibrium state. If the equality holds,

$$(\delta^2 E)_{S,V,n} = 0,$$

the stability is undetermined and one must examine higher order variations. If

$$(\delta^2 E)_{S,V,n} < 0,$$

the system is not stable and the slightest fluctuation or disturbance will cause the system to change macroscopically. (For the purpose of visualization, the reader might consider these remarks in light of the

Fig. 2.3. Arbitrary composite system with two compartments.

analogy with a potential energy function containing both maxima and minima.)

As an example, consider a composite system with two compartments, as pictured in Fig. 2.3, and the fluctuations

$$\delta S = 0 = \delta S^{(1)} + \delta S^{(2)}$$

and

$$\delta V^{(1)} = \delta V^{(2)} = \delta n^{(1)} = \delta n^{(2)} = 0.$$

Then

$$\delta^2 E = (\delta^2 E)^{(1)} + (\delta^2 E)^{(2)}$$

$$= \frac{1}{2} \left(\frac{\partial^2 E}{\partial S^2} \right)^{(1)}_{V,n} (\delta S^{(1)})^2 + \frac{1}{2} \left(\frac{\partial^2 E}{\partial S^2} \right)^{(2)}_{V,n} (\delta S^{(2)})^2,$$

where the superscripts (1) and (2) on the derivatives indicate that the derivatives are to be evaluated at equilibrium for subsystems (1) and (2), respectively. Since $\delta S^{(1)} = -\delta S^{(2)}$, and

$$(\partial^2 E / \partial S^2)_{V,n} = (\partial T / \partial S)_{V,n} = T / C_v,$$

we have

$$(\delta^2 E)_{S,V,n} = \tfrac{1}{2} (\delta S^{(1)})^2 \left[\frac{T^{(1)}}{C_v^{(1)}} + \frac{T^{(2)}}{C_v^{(2)}} \right]$$

$$= \tfrac{1}{2} (\delta S^{(1)})^2 T [1/C_v^{(1)} + 1/C_v^{(2)}],$$

where the second equality follows from $T^{(1)} = T^{(2)} = T$ at equilibrium. By applying $(\delta^2 E)_{S,V,n} \geq 0$, we therefore find

$$T [1/C_v^{(1)} + 1/C_v^{(2)}] \geq 0,$$

or since the division into subsystems can be arbitrary, this result implies

$$T / C_v \geq 0 \quad \text{or} \quad C_v \geq 0.$$

Exercise 2.4 Derive an analogous result for C_p. Here, it may be useful to consider the enthalpy, $H = E + pV$.

Thus, a stable system will have a positive C_v. If it did not, imagine the consequences: Suppose two subsystems (1) and (2) were in thermal contact but not in equilibrium, $T^{(1)} \neq T^{(2)}$. The gradient in T will cause heat flow, and the flow will be from high T to low T. But if $C_v < 0$, the direction of heat flow would cause the gradient in T to grow, and the system would not equilibrate. This illustrates the physical content of stability criteria. If they are obeyed, the spontaneous process induced by a deviation from equilibrium will be in a direction to restore equilibrium.

As another example, look at the Helmholtz free energy

$$(\Delta A)_{T,V,n} > 0, \qquad (\delta A)_{T,V,n} = 0, \qquad (\delta^2 A)_{T,V,n} \geqslant 0.$$

One should note that it is not permissible to consider fluctuations in T since these variational theorems refer to experiments in which internal constraints vary internal extensive variables keeping the total extensive variables fixed. T is intensive, however, and it makes no sense to consider a repartitioning of an intensive variable.

The theorems are applicable to the variations

$$\delta V = 0 = \delta V^{(1)} + \delta V^{(2)}$$

and

$$\delta n^{(1)} = \delta n^{(2)} = 0,$$

where we are once again considering the composite system with two subsystems. The second-order variation in A is

$$(\delta^2 A)_{T,V,n} = \tfrac{1}{2}(\delta V^{(1)})^2 \left[\left(\frac{\partial^2 A}{\partial V^2} \right)^{(1)}_{T,n} + \left(\frac{\partial^2 A}{\partial V^2} \right)^{(2)}_{T,n} \right].$$

Since

$$\left(\frac{\partial^2 A}{\partial V^2} \right)_{T,n} = - \left(\frac{\partial p}{\partial V} \right)_{T,n},$$

the positivity of $(\delta^2 A)_{T,V,n}$ implies

$$- \left[\left(\frac{\partial p}{\partial V} \right)^{(1)}_{T,n} + \left(\frac{\partial p}{\partial V} \right)^{(2)}_{T,n} \right] \geqslant 0,$$

and since the division into subsystems can be arbitrary

$$- \left(\frac{\partial p}{\partial V} \right)_{T,n} \geqslant 0$$

or

$$K_T \geqslant 0,$$

where

$$K_T = - \frac{1}{V} \left(\frac{\partial V}{\partial p} \right)_{T,n}$$

is the isothermal compressibility. Thus, if the pressure of a stable system is increased isothermally, its volume will decrease.

Exercise 2.5 Prove an analogous theorem for the adiabatic compressibility,

$$K_S = -\frac{1}{V}\left(\frac{\partial V}{\partial p}\right)_{S,n}.$$

Exercise 2.6 Show that since $K_T > 0$, stability implies that $C_p > C_v$.

As one further illustration of what can be learned from a stability analysis, let us now suppose $(\delta^2 A)_{T,V,n} = 0$. That is, consider the situation for which $(\partial p/\partial V)_{T,n} = (\partial p/\partial v)_T = 0$, where $v = V/n$, when $-\delta V^{(1)} = \delta V^{(2)}$, but $\delta n^{(1)} = \delta n^{(2)} = 0$. Then

$$0 < (\Delta A)_{T,V,n} = (\delta^3 A)_{T,V,n} + (\delta^4 A)_{T,V,n} + \cdots,$$

Thus, by considering arbitrary small displacements, we see that $(\partial p/\partial v)_T = 0$ implies

$$(\delta^3 A)_{T,V,n} \geqslant 0.$$

Thus,

$$0 \leqslant (\delta V^{(1)})^3 \left[\left(\frac{\partial^3 A}{\partial V^3}\right)^{(1)}_{T,n} - \left(\frac{\partial^3 A}{\partial V^3}\right)^{(2)}_{T,n} \right].$$

Since this equation must hold for all small $\delta V^{(1)}$, both positive and negative, the term in square brackets must be zero. Further,

$$\left(\frac{\partial^3 A}{\partial V^3}\right)^{(1)}_{T,n} = -\left(\frac{\partial^2 p}{\partial V^2}\right)^{(1)}_{T,n} = -\left(\frac{\partial^2 p}{\partial v^2}\right)^{(1)}_T \left(\frac{1}{n^{(1)}}\right)^2.$$

Thus

$$\left(\frac{\partial^2 p}{\partial v^2}\right)^{(1)}_T \left(\frac{1}{n^{(1)}}\right)^2 - \left(\frac{\partial^2 p}{\partial v^2}\right)^{(2)}_T \left(\frac{1}{n^{(2)}}\right)^2 = 0,$$

and since the division into subsystems can be arbitrary, we have proven that if $(\partial p/\partial v)_T = 0$, then $(\partial^2 p/\partial v^2)_T = 0$.

Exercise 2.7 Determine the sign of $(\partial^3 p/\partial v^3)_T$ if, for a thermodynamically stable system, $(\partial p/\partial v)_T = 0$. Determine what is known about $(\partial^4 p/\partial v^4)_T$ if $(\partial^3 p/\partial v^3)_T = 0$.

A general rule for stability criteria should now be apparent. Let Φ stand for the internal energy or a Legendre transform of it which is a natural function of the extensive variables $X_1, X_2, \ldots, X_r$, and the intensive variables $I_{r+1}, \ldots, I_n$. Then

$$d\Phi = \sum_{i=1}^{r} I_i \, dX_i - \sum_{j=r+1}^{n} X_j \, dI_j,$$

and the stability criteria are

$$0 \leq \left(\frac{\partial I_i}{\partial X_i}\right)_{X_1,\ldots,X_{i-1},X_{i+1},\ldots,X_r,I_{r+1},\ldots,I_n}.$$

Exercise 2.8 Derive this result.

Thus, for example, the derivatives

$$-\left(\frac{\partial p}{\partial v}\right)_s, \quad \left(\frac{\partial \mu_i}{\partial n_i}\right)_{T,V,n_j}, \quad \text{and} \quad C_p$$

must all be positive (or zero). However, the second law (i.e. stability) says nothing about the sign of

$$\left(\frac{\partial p}{\partial T}\right)_v \quad \text{and} \quad \left(\frac{\partial \mu_j}{\partial n_l}\right)_{T,V,n_{i(\neq l)}}$$

since these are not derivatives of intensive properties with respect to their conjugate variables.

Exercise 2.9 An experimentalist claims to find that a particular gaseous material obeys the conditions

(i) $(\partial p/\partial v)_T < 0$
(ii) $(\partial p/\partial T)_v > 0$
(iii) $(\partial \mu/\partial v)_T < 0$
(iv) $(\partial T/\partial v)_s > 0$

(a) Identify which of these inequalities is guaranteed by stability.
(b) Identify which pair of inequalities is inconsistent with each other and demonstrate why they are inconsistent.

2.3 Application to Phase Equilibria

Suppose v phases are coexisting in equilibrium. At constant T and p, the conditions for equilibrium are

$$\mu_i^{(\alpha)}(T, p, x_1^{(\alpha)}, \ldots, x_{r-1}^{(\alpha)}) = \mu_i^{(\gamma)}(T, p, x_1^{(\gamma)}, \ldots, x_{r-1}^{(\gamma)}),$$

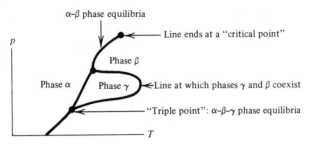

Fig. 2.4. A hypothetical phase diagram.

for $1 \leq \alpha < \gamma \leq v$ and $1 \leq i \leq r$. Here, $x_i^{(\alpha)}$ is the mole fraction of species i in phase α. This relation is an abbreviation for $r(v-1)$ independent equations which couple together $2 + v(r-1)$ different intensive variables (T, p, and the mole fractions for each phase). Hence, the thermodynamic degrees of freedom (the number of independent intensive thermodynamic variables) is

$$f = 2 + v(r-1) - r(v-1)$$
$$= 2 + r - v.$$

This formula is the Gibbs phase rule.

As an illustration, consider a one-component system. Without coexisting phases, there are two degrees of freedom; p and T are a convenient set. The system can exist anywhere in the p-T plane. Three phases coexist at a point, and it is impossible for more than three phases to coexist in a one-component system. Thus, a possible *phase diagram* is illustrated in Fig. 2.4.

The equations that determine the lines on this picture are

$$\mu^{(\alpha)}(p, T) = \mu^{(\beta)}(p, T),$$
$$\mu^{(\alpha)}(p, T) = \mu^{(\gamma)}(p, T),$$

and

$$\mu^{(\beta)}(p, T) = \mu^{(\gamma)}(p, T).$$

For example, the content of the first of these equations is illustrated in Fig. 2.5.

The second law says that at constant T, p, and n, the stable equilibrium state is the one with the lowest Gibbs free energy (which is $n\mu$ for a one-component system). This condition determines which of the two surfaces corresponds to the stable phase on a particular side of the α-β coexistence line.

According to this picture, a *phase transition* is associated with the intersection of Gibbs surfaces. The change in volume in moving from

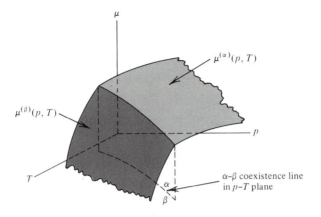

Fig. 2.5. Chemical potential surfaces for two phases.

one surface to the other isothermally is given by the change in

$$(\partial\mu/\partial p)_T = v.$$

The change in entropy associated with the phase transition is given by the change in

$$(\partial\mu/\partial T)_p = -s.$$

If the two surfaces happen to join smoothly to one another, then v and s are continuous during the phase change. When that happens, the transition is called *second order* or *higher order*. A *first-order* transition is one in which, for example, $v(T, p)$ is discontinuous. For a one-component system, a second-order transition can occur at only one point—a critical point. In a two-component system, one can find lines of second-order phase transitions, which are called critical lines.

The p-T coexistence line satisfies a differential equation that is easily derived from the equilibrium condition

$$\mu^{(\alpha)}(T, p) = \mu^{(\beta)}(T, p).$$

Since $d\mu = -s\,dT + v\,dp$,

$$-s^{(\alpha)}\,dT + v^{(\alpha)}\,dp = -s^{(\beta)}\,dT + v^{(\beta)}\,dp,$$

or

$$\frac{dp}{dT} = \frac{\Delta s(T)}{\Delta v(T)},$$

where $\Delta s(T) = s^{(\alpha)}(T, p) - s^{(\beta)}(T, p)$ and $\Delta v(T) = v^{(\alpha)}(T, p) - v^{(\beta)}(T, p)$, at a value of T and p for which phases α and β are at equilibrium. This equation is known as the *Clausius–Clapeyron*

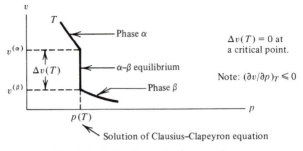

$\Delta v(T) = 0$ at
a critical point.

Note: $(\partial v/\partial p)_T \leqslant 0$

Fig. 2.6. An isotherm in the p-v plane.

equation. Notice that the right-hand side is ill-defined at a second-order phase transition.

Exercise 2.10 Derive the analogous differential equation that describes the α-β coexistence line in the μ-T plane.

Exercise 2.11 Use the Clausius–Clapeyron equation to determine the slope of dp/dT for water-ice I equilibrium and explain why you can skate on ice but not on solid argon.

Another way to view phase equilibria is to look at a thermodynamic plane in which one axis is an intensive field and the other axis is the conjugate variable to this field. For example, consider the p-v plane for a one-component system. In Fig. 2.6, the quantity $v^{(\alpha)}$ is the volume per mole of pure phase α when it is in equilibrium with phase β at a temperature T. $v^{(\beta)}$ has a similar definition. Notice how $v(T, p)$ is discontinuous (i.e., the system suffers a first-order phase transition) as we pass isothermally from a pressure just below $p(T)$ to one just above $p(T)$. The equations that determine $v^{(\alpha)}(T)$ and $v^{(\beta)}(T)$ are

$$p(T) = p^{(\alpha)}(T, v^{(\alpha)}) = p^{(\beta)}(T, v^{(\beta)})$$

and

$$\mu(T) = \mu^{(\alpha)}(T, v^{(\alpha)}) = \mu^{(\beta)}(T, v^{(\beta)}),$$

where $\mu^{(\alpha)}(T, v)$ and $p^{(\alpha)}(T, v)$ are the chemical potential and pressure, respectively, as a function of T and V for phase α.

Here is a puzzle to think about: For water near 1 atm pressure and 0°C temperature, the solid phase, ice I, has a larger volume per mole than the liquid. Does this mean that the 0°C isotherm appears as

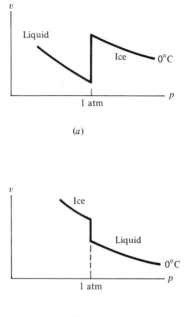

(a)

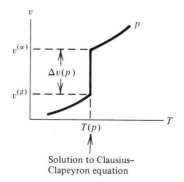

(b)

Fig. 2.7a and b. Which one might be an isotherm for water?

pictured in Fig. 2.7a? Wouldn't this behavior violate stability?
Perhaps Fig. 2.7b is correct.

Exercise 2.12 What is the correct answer to this puzzle?

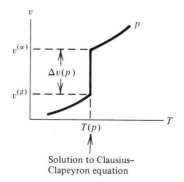

Solution to Clausius–
Clapeyron equation

Fig. 2.8. Isobar in the v-T plane.

We can also look at the v-T plane. For many systems, the picture looks like that shown in Fig. 2.8. At times this representation is very informative. But sometimes it becomes difficult to use because v and T are not conjugate, and as a result, $v(T, p)$ is not necessarily a monotonic function of T.

Exercise 2.13* Draw a family of curves like that shown in Fig. 2.8 for the v-T plane for water and ice I near 1°C and 1 atm.

The solution of the two coupled equations given above for $v^{(\alpha)}(T)$ and $v^{(\beta)}(T)$ can be given a geometric interpretation, which is called a *Maxwell construction*. Let $a = A/n$, and consider a graph of a vs. v at constant T. If a phase transition occurs, there will be a region corresponding to pure phase α with $a = a^{(\alpha)}$, a region corresponding to pure phase β with $a = a^{(\beta)}$, and something in between. The a vs. v isotherm appears as shown in Fig. 2.9. To prove that a double tangent line connects $a^{(\alpha)}$ to $a^{(\beta)}$ at the volumes $v^{(\alpha)}$ and $v^{(\beta)}$, respectively, note that

$$\left(\frac{\partial a}{\partial v}\right)_T = -p.$$

Hence, the equilibrium condition

$$p(T) = p^{(\alpha)}(T, v^{(\alpha)}) = p^{(\beta)}(T, v^{(\beta)})$$

implies that the slope at $v = v^{(\alpha)}$ is the same as that at $v = v^{(\beta)}$. The common tangent construction then gives

$$a(T, v^{(\alpha)}) - a(T, v^{(\beta)}) = -p(T)[v^{(\alpha)}(T) - v^{(\beta)}(T)]$$

or

$$(a + pv)^{(\alpha)} = (a + pv)^{(\beta)},$$

which is the equilibrium condition, $\mu^{(\alpha)} = \mu^{(\beta)}$.

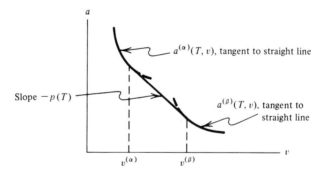

Fig. 2.9. The Helmholtz free energy per mole on an isotherm.

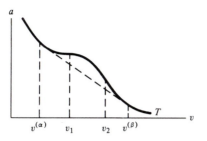

Fig. 2.10. A questionable Helmholz free energy per mole on an isotherm.

Finally, since p is fixed by T when two phases are in equilibrium, the double tangent line drawn between $v^{(\alpha)}$ and $v^{(\beta)}$ is the free energy per mole in the two-phase region ($v^{(\alpha)} < v < v^{(\beta)}$). Hence

$$\left(\frac{A}{n}\right)_{\substack{\text{two-phase} \\ \text{region}}} = a^{(\alpha)} + \frac{v - v^{(\alpha)}}{v^{(\beta)} - v^{(\alpha)}}[a^{(\beta)} - a^{(\alpha)}]$$

$$= a^{(\alpha)}\left(\frac{v^{(\beta)} - v}{v^{(\beta)} - v^{(\alpha)}}\right) + a^{(\beta)}\left(\frac{v - v^{(\alpha)}}{v^{(\beta)} - v^{(\alpha)}}\right),$$

where

$$a^{(\alpha)} = a(T, v^{(\alpha)}(T))$$

and

$$a^{(\beta)} = a(T, v^{(\beta)}(T))$$

are the free energies per mole of the two pure phases when they are in equilibrium with each other.

Exercise 2.14 Draw the analogous graph for (A/V) vs. v^{-1} for an isotherm on which a phase transition occurs.

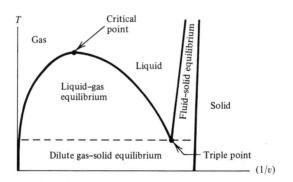

Fig. 2.11. Phase diagram for a simple material.

Some approximate thermodynamic theories have free energies that look like the one pictured here in Fig. 2.10, which cannot be right because in the region $v_1 < v < v_2$ stability is violated—that is,

$$\left(\frac{\partial^2 a}{\partial v^2}\right)_T = -\left(\frac{\partial p}{\partial v}\right)_T < 0,$$

for v between v_1 and v_2. One assumes for these theories that the instability is bridged by a phase transition located by a Maxwell construction (the dashed line).

Exercise 2.15 The van der Waals equation of state is

$$p/RT = \rho/(1 - b\rho) - a\rho^2/RT,$$

where R, b, and a are positive constants and $\rho = n/V$. Show that below a certain temperature the van der Waals equation of state implies a free energy that is unstable for some densities. (Restrict analysis to $\rho < b^{-1}$.)

The locus of points formed from $v^{(\alpha)}$ and $v^{(\beta)}$ at different temperatures gives a coexistence curve. For example, in a one-component simple fluid like argon the phase diagram looks like the diagram pictured in Fig. 2.11.

Exercise 2.16* Draw the analogous curves for water.

If an approximate theory is used in which an instability is associated with a phase transition, the locus of points surrounding the unstable region is called the *spinodal*. The spinodal must be enveloped by the coexistence curve. For example, the van der Waals equation yields a diagram like the one pictured in Fig. 2.12.

2.4 Plane Interfaces

If two phases are in equilibrium, there is a surface or interface of material between them. Let us now focus attention on this interface. (See Fig. 2.13.)

The density profile near the interface is sketched in Fig. 2.14. In this figure, $\rho(z)$ is the number of moles (or molecules) per unit volume of a particular species, z_d is the (arbitrary) location of the *dividing surface,* and w is the width of the interface (typically a few molecular diameters).

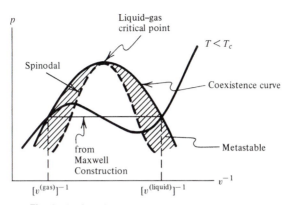

Fig. 2.12. Coexistence curves and spinodals.

Since E is extensive,

$$E = E^{(\alpha)} + E^{(\beta)} + E^{(s)},$$

where $E^{(s)}$ is the energy of the interface. This surface energy should depend upon the surface area, σ. Let

$$\gamma = (\partial E / \partial \sigma)_{S,V,n} \geq 0.$$

Then

$$dE = T \, dS - p \, dV + \mu \, dn + \gamma \, d\sigma.$$

The property γ is called the *surface tension*. It is intensive by virtue of its definition. It also should be positive. If not, lower energy states would be obtained by making the phase boundary more irregular since the irregularity will increase the surface area. Therefore, a negative surface tension would drive the system to a terminal state in which the surface was spread over the whole system. The boundary between two phases would then cease to exist, and there would be no "surface."

Since the interface exists when there is two-phase equilibria, the Gibbs phase rule tells us that γ is determined by r intensive variables where r is the number of components. For one component, T suffices. Clearly, E is first order homogeneous in S, V, n, and σ.

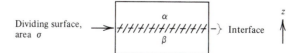

Fig. 2.13. Hypothetical interface between two phases α and β.

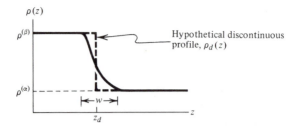

Fig. 2.14. Density profile.

Thus,

$$E = TS - pV + \mu n + \gamma \sigma.$$

Note that while E is of the order of N = total number of molecules, $\gamma \sigma$ is of the order of $N^{2/3}$. Thus, under normal circumstances, the surface energy has a negligible effect upon bulk properties.

Exercise 2.17* You might be concerned that if we double σ while doubling S, V, and n, we must keep the geometry of the system as a "slab" of fixed thickness. This means that there will be surface interaction energies with the walls that are also proportional to σ. How do we separate these energies from $\gamma \sigma$? Can you think of procedures to do this? Consider two sorts: (i) Make measurements of containers filled with pure vapor, pure liquid, and a mixture of the two. (ii) Employ a container with periodic boundary conditions. The latter might be difficult in practice, but could be feasible in the world one can simulate with computers (see Chapter 6).

Imagine a hypothetical phase that would result if phase α maintained its bulk properties right up to the mathematical dividing surface. For such a phase

$$dE^{(\alpha)} = T\, dS^{(\alpha)} - p\, dV^{(\alpha)} + \mu\, dn^{(\alpha)},$$
$$E^{(\alpha)} = TS^{(\alpha)} - pV^{(\alpha)} + \mu n^{(\alpha)}.$$

Similarly,

$$dE^{(\beta)} = T\, dS^{(\beta)} - p\, dV^{(\beta)} + \mu\, dn^{(\beta)},$$
$$E^{(\beta)} = TS^{(\beta)} - pV^{(\beta)} + \mu n^{(\beta)}.$$

Of course, $V^{(\alpha)} + V^{(\beta)} = V$, the total volume of the system. For any extensive property, X, we can define surface excess extensive property, $X^{(s)}$, by

$$X^{(s)} = X - X^{(\alpha)} - X^{(\beta)},$$

since X is well defined, and $X^{(\alpha)}$ and $X^{(\beta)}$ are defined in terms of bulk properties once the location of the dividing surface is specified. Obviously, $V^{(s)} = 0$. What about $n^{(s)}$? Clearly,

$$n^{(\beta)} = \int_{-\infty}^{z_d} dz \, \rho_d(z), \qquad n^{(\alpha)} = \int_{z_d}^{\infty} dz \, \rho_d(z),$$

where $\rho_d(z)$ is the hypothetical discontinuous density profile pictured in Fig. 2.14. Thus,

$$n^{(s)} = \int_{-\infty}^{\infty} dz [\rho(z) - \rho_d(z)].$$

By viewing the figure showing $\rho(z)$ and $\rho_d(z)$, it is seen that there exists a choice of z_d for which $n^{(s)} = 0$. That particular choice is called the *Gibbs dividing surface*. That is, the z_d that solves

$$n^{(s)}(z_d) = 0$$

is the location of the Gibbs surface.

The energy differential for $E^{(s)}$ is

$$dE^{(s)} = dE - dE^{(\alpha)} - dE^{(\beta)}$$
$$= T \, dS^{(s)} + \gamma \, d\sigma + \mu \, dn^{(s)},$$

where we have noted that $dV = dV^{(\alpha)} + dV^{(\beta)}$. If the Gibbs surface is chosen as the location of z_d,

$$dE^{(s)} = T \, dS^{(s)} + \gamma \, d\sigma.$$

With this choice,

$$E^{(s)} = TS^{(s)} + \gamma\sigma,$$

so that

$$\gamma = (E^{(s)} - TS^{(s)})/\sigma.$$

In other words, the surface tension is the surface Helmholtz free energy per unit area. By virtue of its definition, γ also plays the role of the force constant in the restoring force that inhibits the growth of surface area [i.e., recall $(dW)_{\text{surface}} = \gamma \, d\sigma$].

In the choice of Gibbs surface previously discussed, we removed the term in $E^{(s)}$ that depends upon mole number. The energetics of the surface is a result of the repartitioning of the bulk phases that occurs if the surface is altered. The absence of any dependence on mole number, however, can only be accomplished for one species. To develop the surface energy expressions for a mixture, we begin with our expression for $dE^{(s)}$ before the Gibbs surface is employed. That is,

$$dE^{(s)} = T \, dS^{(s)} + \gamma \, d\sigma + \sum_{i=1}^{r} \mu_i \, dn_i^{(s)}.$$

Let component 1 be the solvent and the rest the solutes, and let us choose z_d to make $n_1^{(s)} = 0$. Then with this choice of Gibbs surface

$$dE^{(s)} = T \, dS^{(s)} + \gamma \, d\sigma + \sum_{i=2}^{r} \mu_i \, dn_i^{(s)}.$$

For $r = 2$,

$$dE^{(s)} = T \, dS^{(s)} + \gamma \, d\sigma + \mu_2 \, dn_2^{(s)},$$

which implies

$$E^{(s)} = TS + \gamma\sigma + \mu_2 n_2^{(s)}.$$

Hence, we have a Gibbs–Duhem-like equation

$$\sigma \, d\gamma = -S^{(s)} \, dT - n_2^{(s)} \, d\mu_2.$$

At constant T this yields

$$\sigma \, d\gamma = -n_2^{(s)} \, d\mu_2.$$

The solution to this equation is called the *Gibbs adsorption isotherm*. Rearranging the equation, we get

$$(n_2^{(s)}/\sigma) = -(\partial\gamma/\partial\mu_2)_T$$
$$= -(\partial\gamma/\partial\rho_2)_T (\partial\rho_2/\partial\mu_2)_T.$$

Since $(\partial\mu_2/\partial\rho_2)_T > 0$ by stability, this equation says that surface tension decreases when a solute accumulates on a surface.

Exercise 2.18 Put some water in a bowl. Sprinkle pepper on its surface. Touch a bar of soap to the surface in the middle of the bowl. What happened? Touch the surface with the soap again, and again. What happened and why? [*Hint*: Consider the formula for the Gibbs adsorption isotherm.]

Before ending this section on interfaces and surface tension, a few qualitative remarks are in order. First, let us consider two immiscible liquids, perhaps an oil and water equilibrium. In a gravitational field, the heavier phase will fall to the bottom of the container, and a planar interface will form between the two liquids. In the absence of gravity, we can imagine that one of the fluids forms a spherical drop of liquid surrounded by the other species of fluid. We suspect that the drop would be spherical since this shape will minimize the surface area of the interface and thereby minimize the contribution to the free energy from the surface tension. Deformation of the surface will lead to higher curvature, larger surface area, and, therefore, higher surface free energy. The restoring force that opposes this deformation is proportional to the surface tension. When this tension vanishes, the deformations are not hindered; the interface will then fluctuate wildly, the drop will become amoeboid and eventually disintegrate. In other words, the two fluids will mix. This type of behavior can be observed by adding a third component to an oil-water mixture (perhaps a surfactant*) that would lower the surface tension between the oil and water phases.

In some cases, the mixing of the phases is associated with the formation of small assemblies such as micelles. A typical micelle is an aggregate involving roughly 100 surfactant molecules. Here, one imagines that the charged or polar head groups lie on a surface surrounding the hydrophobic tails and oil thus inhibiting contact between the water and oil. The surface tension, however, is relatively low so that the shapes of these assemblies undoubtedly fluctuate appreciably. Further, if one considers small enough systems at a truly molecular level, fluctuations are almost always significant. Statistical mechanics is the subject that describes the nature of these fluctuations, our next topic in this book.

In Chapter 6 we study the phase equilibrium of a small system by a numerical Monte Carlo simulation. A glance at the results of those simulations provides ample evidence of the importance of fluctuations at interfaces. Their importance must be considered carefully in any microscopic formulation of interfacial phenomena. Indeed, the reader may now wonder whether an interface with an intrinsic width as pictured at the beginning of this section is actually well defined at a microscopic level. It's an important and puzzling issue worth thinking about.

Additional Exercises

2.19. Show that

$$(K_s/K_T) = C_v/C_p,$$

where K_s and K_T are the adiabatic and isothermal compressibilities, respectively, and C_v and C_p are the constant volume and constant pressure heat capacities, respectively. Prove that for any stable system

$$K_s < K_T.$$

---

* Surfactant molecules are *amphiphiles*. That is, they possess a hydrophilic (charged or polar) head group, and a hydrophobic (oil-like) tail.

2.20. (a) It is easily verified that a rubber band heats up when it is stretched adiabatically. Given this fact, determine whether a rubber band will contract or expand when it is cooled at constant tension.

(b) The same amount of heat flows into two identical rubber bands, but one is held at constant tension and the other at constant length. Which has the largest increase in temperature?

2.21. (a) For many systems, the work differential associated with a magnetic field of strength, H, is either

$$M\,dH \quad \text{or} \quad H\,dM,$$

where M is the net magnetization in the direction of the magnetic field. Which of these is correct? For a stable system, determine the signs of the isothermal and adiabatic susceptibilities

$$\chi_T = (\partial M/\partial H)_{T,n,V},$$
$$\chi_S = (\partial M/\partial H)_{S,n,V}.$$

Determine the sign of $\chi_T - \chi_S$.

(b) For most paramagnetic substances, the magnetization at constant H is a decreasing function of T. Given this fact, determine what happens to the temperature of a paramagnetic material when it is adiabatically demagnetized. That is, determine the sign of $(\partial T/\partial H)_{S,V,n}$.

2.22. Consider a one-component system when two phases, α and β, are in equilibrium. Show that the locus of points at which this equilibrium occurs in the μ-T plane is given by the solution to the differential equation

$$\frac{d\mu}{dT} = \frac{s^{(\beta)}v^{(\alpha)} - s^{(\alpha)}v^{(\beta)}}{v^{(\beta)} - v^{(\alpha)}},$$

where $s^{(\alpha)}$ and $v^{(\alpha)}$ are the entropy and volume, respectively, per mole of phase α when it is in equilibrium with phase β. Why is $d\mu/dT$ a total derivative and not a partial? Suppose the system contains two species. Derive the appropriate generalization of the equation above. That is, determine $(\partial \mu_1/\partial T)_{x_1}$ at two-phase equilibrium, where x_1 is the mole fraction of component 1.

2.23. It is found that when stretched to a certain length, a particular spring breaks. Before the spring breaks (i.e., at small lengths), the free energy of the spring is given by

$$\frac{A}{M} = \tfrac{1}{2}kx^2,$$

where $A = E - TS$, M is the mass of the spring, and x is its length per unit mass. After breaking (i.e., at larger lengths)

$$\frac{A}{M} = \tfrac{1}{2}h(x - x_0)^2 + c.$$

In these equations, k, h, x_0, and c are all independent of x, but do depend on T. Furthermore, $k > h$, $c > 0$, and $x_0 > 0$ for all T.

(a) Determine the equations of state

$$f = \text{tension} = f(T, x),$$

for the spring at small and long lengths.
(b) Similarly, determine the chemical potentials

$$\mu = (\partial A / \partial M)_{T,L},$$

where L is the total length of the spring.
(c) Show that

$$\mu = \frac{A}{M} - fx.$$

(d) Find the force that at given temperature will break the spring.
(e) Determine the discontinuous change in x when the spring breaks.

2.24. A hypothetical experimentalist measures the hypothetical equation of state for a substance near the liquid-solid phase transition. He finds that over a limited range of temperatures and densities, the liquid phase can be characterized by the following formula for the Helmholtz free energy per unit volume:

$$A^{(l)}/V = \tfrac{1}{2}a(T)\rho^2,$$

where the superscript "l" denotes "liquid," $\rho = n/V$ is the molar density, and $a(T)$ is a function of the temperature,

$$a(T) = \alpha/T, \qquad \alpha = \text{constant}.$$

Similarly, in the solid phase he finds

$$A^{(s)}/V = \tfrac{1}{3}b(T)\rho^3,$$

with

$$b(T) = \beta/T, \qquad \beta = \text{constant}.$$

At a given temperature, the pressure of the liquid can be adjusted to a particular pressure, p_s, at which point the liquid freezes. Just before solidification, the density is $\rho^{(l)}$, while just after it is $\rho^{(s)}$.

(a) Determine $\rho^{(l)}$ and $\rho^{(s)}$ as functions of the temperature.
(b) Determine p_s as function of temperature.
(c) Determine the change in entropy per mole during solidification.
(d) By using the Clausius–Clapeyron equation and the results of parts (a) and (c), determine the slope of (dp/dT) at solidification. Does your result agree with what you would predict from the solution to part (b)?

2.25. The van der Waals equation is

$$p/RT = \rho/(1 - b\rho) - a\rho^2/RT,$$

where $\rho = n/V$ and a and b are constants. Show that there is a region in the T-ρ plane in which this equation violates stability. Determine the boundary of this region; that is, find the spinodal. Prove that a Maxwell construction will yield a liquid-gas coexistence curve which will envelop the region of instability.

2.26. When a particular one-component material is in phase α, it obeys equation of state

$$\beta p = a + b\beta\mu,$$

where $\beta = 1/T$, and a and b are positive functions of β. When it is in phase γ,

$$\beta p = c + d(\beta\mu)^2,$$

where c and d are positive functions of β, $d > b$, and $c < a$. Determine the density change that occurs when the material suffers a phase transformation from phase α to phase γ. What is the pressure at which the transition occurs?

Bibliography

Treatments of equilibria and stability criteria close to that presented in this chapter are given in

J. G. Kirkwood and I. Oppenheim, *Chemical Thermodynamics* (McGraw-Hill, N.Y., 1960),

and

J. Beatte and I. Oppenheim, *Thermodynamics* (Elsevier Scientific, N.Y., 1979).

The thermodynamic treatments of phase equilibria, interfaces, and surface tension are found in many texts. Two texts with useful and brief discussions are

T. L. Hill, *Thermodynamics for Chemists and Biologists* (Addison-Wesley, Reading, Mass., 1968),

and

E. M. Lifshitz and L. P. Pitaevskii, *Statistical Physics, 3rd ed., Part* 1 (Pergamon, N.Y., 1980).

This last one is a revision of the L. D. Landau and E. M. Lifshitz classic, *Statistical Physics* (Pergamon, N.Y., 1958). The monograph devoted to interfacial phenomena of fluids is

J. S. Rowlinson and B. Widom, *Molecular Theory of Capillarity* (Oxford University Press, Oxford, 1982).

Surface tension provides the mechanism for many dramatic phenomena. Some colorful yet simple demonstrations that exploit the fact that oil and water don't mix (the *hydrophobic effect*) are described in the articles

J. Walker, *Scientific American* **249**, 164 (1982); and F. Sebba in *Colloidal Dispersions and Micellar Behavior* (ACS Symposium Series No. 9, 1975).

CHAPTER 3

Statistical Mechanics

We now turn our attention to the molecular foundation of thermo-dynamics, or more generally, the answer to the following question: If particles (atoms, molecules, or electrons and nuclei, ...) obey certain microscopic laws with specified interparticle interactions, what are the observable properties of a system containing a very large number of such particles? That is, we want to discuss the relationship between the microscopic dynamics or fluctuations (as governed by Schrödinger's equation or Newton's laws of motion) and the observed properties of a large system (such as the heat capacity or equation of state).

The task of solving the equations of motion for a many-body system (say N = number of particles $\sim 10^{23}$) is so complicated that even modern day computers find the problem intractable. (Though scientists do use computers to follow the motion of thousands of particles for times often long enough to simulate condensed phases for times of the order of 10^{-10} or 10^{-9} sec.) At first you might think that as the number of particles increases, the complexity and obscurity of the properties of a mechanical system should increase tremendously, and that you would be unable to find any regularity in the behavior of a macroscopic body. But as you know from thermodynamics, large systems are, in a sense, quite orderly. An example is the fact that at thermodynamic equilibrium one can characterize observations of a macroscopic system with only a handful of variables. The attitude we shall take is that these distinctive regularities are consequences of statistical laws governing the behavior of systems composed of very many particles. We will thereby avoid the need to directly evaluate the precise N-particle dynamics, and assume that probability statistics provides the correct description of what we see during a macroscopic measurement.

The word "measurement" is important in these remarks. If we imagined, for example, observing the time evolution of one particular particle in a many-body system, its energy, its momentum, and its position would all fluctuate widely, and the precise behavior of any of these properties would change drastically with the application of the slightest perturbation. One cannot imagine a reproducible measurement of such chaotic properties since even the act of observation involves a perturbation. Further, to reproduce the precise time evolution of a many-body system, one must specify at some initial time a macroscopic number ($\sim 10^{23}$) of variables. These variables are initial coordinates and momenta of all the particles if they are classical, or an equally cumbersome list of numbers if they are quantal. If we would fail to list just one of these 10^{23} variables, the time evolution of the system would no longer be deterministic, and an observation that depended upon the precise time evolution would no longer be reproducible. It is beyond our capacity to control 10^{23} variables. As a result, we confine our attention to simpler properties, those controlled by only a few variables. In some areas of physical and biological science, it might not be easy to identify those variables. But as a philosophical point, scientists approach most observations with an eye to discovering which small number of variables guarantees the reproducibility of phenomena.

The use of statistics for reproducible phenomena does not imply that our description will be entirely undeterministic or vague. To the contrary, we will be able to predict that the observed values of many physical quantities remain practically constant and equal to their average values, and only very rarely show any detectable deviations. (For example, if one isolates a small volume of gas containing, say, only 0.01 moles of gas, then the average relative deviation of the energy of this quantity from its mean value is $\sim 10^{-11}$. The probability of finding in a single measurement a relative deviation of 10^{-6} is $\sim 10^{-3 \times 10^{15}}$.) As a rough rule of thumb: If an observable of a many particle system can be specified by a small number of other macroscopic properties, we assume that the observable can be described with statistical mechanics. For this reason, statistical mechanics is often illustrated by applying it to equilibrium thermodynamic quantities.

3.1 The Statistical Method and Ensembles

While it is not possible in practice, let us imagine that we could observe a many-body system in a particular microscopic state. Its characterization would require an enormous number of variables. For example, suppose the system was quantal obeying Schrödinger's equation

$$i\hbar \frac{\partial}{\partial t} |\psi\rangle = \mathcal{H} |\psi\rangle.$$

Here, as always, $2\pi\hbar$ is Planck's constant, $\mathcal{H}$ is the Hamiltonian operating on the state vector $|\psi\rangle$, and t is the time. To specify the state $|\psi\rangle$ at a particular time, we need a number of variables of the order of N, the number of particles in the system.

Consider, for example, stationary solutions

$$\mathcal{H} |\psi_v\rangle = E_v |\psi_v\rangle,$$

and some simple and familiar quantum mechanical systems such as the hydrogen atom, or non-interacting particles in a box. The index v is then the collection of $D \cdot N$ quantum numbers, where D is the dimensionality.

Once the initial state is specified, if it could be, the state at all future times is determined by the time integration of Schrödinger's equation. The analogous statement for classical systems considers points in *phase space*

$$(r^N, p^N) \equiv (\mathbf{r}_1, \mathbf{r}_2, \ldots, \mathbf{r}_N; \mathbf{p}_1, \ldots, \mathbf{p}_N),$$

where $\mathbf{r}_i$ and $\mathbf{p}_i$ are the coordinates and conjugate momenta, respectively, for particle i. Points in phase space characterize completely the mechanical (i.e., microscopic) state of a classical system, and flow in this space is determined by the time integration of Newton's equation of motion, $F = ma$, with the initial phase space point providing the initial conditions.

Exercise 3.1 Write down the differential equations corresponding to Newton's laws when the total potential energy is the function $U(\mathbf{r}_1, \mathbf{r}_2, \ldots, \mathbf{r}_N)$.

Now try to think about this time evolution—the trajectory—of a many-body system. As illustrated in Fig. 3.1, we might picture the evolution as a line in "state space" (phase space in the classical case, or Hilbert space spanned by all the state vectors $|\psi\rangle$ in the quantal case). In preparing the system for this trajectory a certain small number of variables is controlled. For example, we might fix the total energy, E, the total number of particles, N, and the volume, V. These constraints cause the trajectory to move on a "surface" of state space—though the dimensionality of the surface is still enormously high.

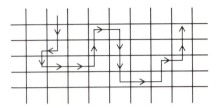

Fig. 3.1. Trajectory in state space with each box representing a different state.

A basic concept in statistical mechanics is that if we wait long enough, the system will eventually flow through (or arbitrarily close to) all the microscopic states consistent with the constraints we have imposed to control the system. Suppose this is the case, and imagine that the system is constantly flowing through the state space as we perform a multitude $\mathcal{N}$ of independent measurements on the system. The observed value ascertained from these measurements for some property G is

$$G_{\text{obs}} = \frac{1}{\mathcal{N}} \sum_{a=1}^{\mathcal{N}} G_a,$$

where G_a is the value during the ath measurement whose time duration is very short—so short, in fact, that during the ath measurement the system can be considered to be in only one microscopic state. Then we can partition the sum as

$$G_{\text{obs}} = \sum_{v} \left[\frac{1}{\mathcal{N}} \left(\begin{matrix} \text{number of times state } v \text{ is} \\ \text{observed in the } \mathcal{N} \text{ observations} \end{matrix} \right) \right] G_v,$$

where $G_v = \langle v | G | v \rangle$ is the expectation value for G when the system is in state v. The term in square brackets is the probability or weight for finding the system during the course of the measurements in state v. Remember, we believe that after a long enough time, all states are visited. We give the probability or fraction of time spent in state v the symbol P_v and write

$$G_{\text{obs}} = \sum_{v} P_v G_v \equiv \langle G \rangle.$$

The averaging operation (i.e., the weighted summation over G_v) indicated by the pointed brackets, $\langle G \rangle$, is called an *ensemble average*. An "ensemble" is the assembly of all possible microstates—all states consistent with the constraints with which we characterize the system macroscopically. For example, the *microcanonical ensemble* is the assembly of all states with fixed total energy E, and fixed size (usually specified by number of molecules, N, and volume, V). The *canonical ensemble,* another example, considers all states with fixed size, but the energy can fluctuate. The former is appropriate to a closed isolated system; the latter is appropriate for a closed system in contact with a heat bath. There will be much more said about these ensembles later.

The idea that we observe the ensemble average, $\langle G \rangle$, arises from the view in which measurements are performed over a long time, and that due to the flow of the system through state space, the time average is the same as the ensemble average. The equivalence of a time average and an ensemble average, while sounding reasonable, is not at all trivial. Dynamical systems that obey this equivalence are said to be *ergodic.* It is difficult, in general, to establish the principle of ergodicity, though we believe it holds for all many-body systems encountered in nature. (It is often true for very small systems too, such as polyatomic molecules. Indeed, the basis of the standard theories of unimolecular kinetics rests on the assumed ergodic nature of intramolecular dynamics.)

Exercise 3.2 Give some examples of *non*-ergodic systems. That is, describe systems that do not sample all possible states even after a very long time.

Incidentally, suppose you thought of employing stationary solutions of Schrödinger's equation to specify microscopic states. If truly in a stationary state at some point in time, the system will remain there for all time, and the behavior cannot be ergodic. But in a many-body system, where the spacing between energy levels is so small as to form a continuum, there are always sources of perturbation or randomness (the walls of the container, for example) that make moot the chance of the system ever settling into a stationary state.

The primary assumption of statistical mechanics—that the observed value of a property corresponds to the ensemble average of that property—seems reasonable, therefore, if the observation is carried out over a very long time or if the observation is actually the average over very many independent observations. The two situations are actually the same if "long time" refers to a duration much longer than any *relaxation time* for the system. The idea that the system is chaotic at a molecular level leads to the concept that after some period of time—a relaxation time, τ_{relax}—the system will lose all memory of (i.e., correlation with) its initial conditions. Therefore, if a measurement is performed over a period τ_{measure} that is $\mathcal{N}\tau_{\text{relax}}$, the measurement actually corresponds to $\mathcal{N}$ independent observations.

In practice, we often consider measurements on macroscopic systems that are performed for rather short periods of time, and the concept of ensemble averages is applicable for these situations, too. This can be understood by imagining a division of the observed macroscopic system into an assembly of many macroscopic subsystems. If the subsystems are large enough, we expect that the precise molecular behavior in one subsystem is uncorrelated with that in any of the neighboring subsystems. The distance across one of these subsystems is then said to be much larger than the *correlation length* or *range of correlations*. When subsystems are this large they behave as if they are macroscopic. Under these conditions, one instantaneous measurement of the total macroscopic system is equivalent to many independent measurements of the macroscopic subsystems. The many independent measurements should correspond to an ensemble average.

3.2 Microcanonical Ensemble and the Rational Foundation of Thermodynamics

The basic idea of statistical mechanics is, therefore, that during a measurement, every microscopic state or fluctuation that is possible does in fact occur, and observed properties are actually the averages from all the microscopic states. To quantify this idea, we need to know something about the probability or distribution of the various microscopic states. This information is obtained from an assumption about the behavior of many-body systems:

> For an isolated system with fixed total energy E, and fixed size (perhaps specified by the volume V and numbers of particles $N_1, N_2, \dots$) all microscopic states are equally likely at thermodynamic equilibrium.

In other words, the macroscopic equilibrium state corresponds to the most random situation—the distribution of microscopic states with the same energy and system size is entirely uniform.

Exercise 3.3 List several everyday examples supporting this statistical characterization of the terminal state of a macroscopic system (e.g., the behavior of a drop of ink in a glass of water).

To examine the implications of this reasonable assumption, let us define

$$\Omega(N, V, E) = \text{number of microscopic states with } N \text{ and } V,$$
$$\text{and energy between } E \text{ and } E - \delta E.$$

For notational and perhaps conceptual simplicity, we often omit subscripts and simply write N to refer to the number of particles, and we use the volume V to specify the spatial extent of the system. Our remarks, however, are not confined to one-component three-dimensional systems. The width δE is some energy interval characteristic of the limitation in our ability to specify absolutely precisely the energy of a macroscopic system. If δE was zero, the quantity $\Omega(N, V, E)$ would be a wildly varying discontinuous function, and when it was non-zero, its value would be the degeneracy of the energy level E. With a finite δE, $\Omega(N, V, E)$ is a relatively continuous function for which standard mathematical analysis is permissible. It will turn out that the thermodynamic consequences are extraordinarily insensitive to the size of δE. The reason for the insensitivity, we will see, is that $\Omega(N, V, E)$ is typically such a rapidly increasing function of E, that any choice of $\delta E \leqslant E$ will usually give the same answer for the thermodynamic consequences examined below. Due to this insensitivity, we adopt the shorthand where the symbol δE is not included in our formulas.

For macroscopic systems, energy levels will often be spaced so closely as to approach a continuum. In the continuum limit it can be convenient to adopt the notation

$$\bar{\Omega}(N, V, E) \, dE = \text{number of states with energy}$$
$$\text{between } E \text{ and } E + dE,$$

where $\bar{\Omega}(N, V, E)$, defined by this equation, is called the *density of states*. In the applications we pursue, however, we will have little need to employ this notation.

Exercise 3.4 For a system with discrete energy levels, give a formula for the density of states, $\bar{\Omega}(N, V, E)$. [*Hint:* You will need to use the Dirac delta function.]

According to the statistical assumption, the probability of a macroscopic state v for an equilibrium system is

$$P_v = 1/\Omega(N, V, E)$$

for all states in the ensemble. For states outside the ensemble, for example those for which $E_v \neq E$, P_v is zero. This ensemble, which is appropriate to a system with fixed energy, volume, and particle number—the assembly of all microstates with these constraints—is called a microcanonical ensemble.

We can also consider as a *definition of entropy* the quantity

$$S = k_B \ln \Omega(N, V, E),$$

where k_B is an arbitrary constant. (It's called *Boltzmann's constant* and we shall find that from comparison with experiment that it has the value

$$k_B = 1.380 \times 10^{-16} \text{ erg/deg.})$$

Notice that S defined in this way is extensive since if the total system were composed of two independent subsystems, A and B, with number of states Ω_A and Ω_B separately, then the total number would be $\Omega_A \Omega_B$. That is, $S_{A+B} = k_B \ln (\Omega_A \Omega_B) = S_A + S_B$.

The definition is also consistent with the variational statements of the second law of thermodynamics. To see why, imagine dividing the system with fixed total N, V, and E into two subsystems and constraining the partitioning of N, V, and E to be $N^{(1)}$, $N^{(2)}$; $V^{(1)}$, $V^{(2)}$; and $E^{(1)}$, $E^{(2)}$, respectively. Any specific partitioning is a subset of all the allowed states, and therefore the number of states with this partitioning, $\Omega(N, V, E; \text{internal constraint})$ is less than the total number $\Omega(N, V, E)$. As a result,

$$S(N, V, E) > S(N, V, E; \text{internal constraint}).$$

This inequality is the second law, and we now see its statistical meaning: the maximization of entropy coinciding with the attainment of equilibrium corresponds to the maximization of disorder or molecular randomness. The greater the microscopic disorder, the larger the entropy.

The temperature T is determined from the derivative $(\partial S / \partial E)_{N,V} = 1/T$. Therefore,

$$\beta = (k_B T)^{-1} = (\partial \ln \Omega / \partial E)_{N,V}.$$

The thermodynamic condition that temperature is positive requires that $\Omega(N, V, E)$ be a monotonic increasing function of E. For macroscopic systems encountered in nature, this will always be the case.

Before accepting this fact as an obvious one, however, consider the following puzzle: Suppose a system of N non-interacting spins in a magnetic field H has the energy

$$-\sum_{j=1}^{N} \mu_j H, \qquad \mu_j = \pm \mu.$$

In the ground state, all the spins are lined up with the field, and $\Omega = 1$. In a first excited state, one spin is flipped and $\Omega = N$. The next excitation has two spins flipped and $\Omega = N(N-1)/2$. Everything looks fine until we realize that the degeneracy of the most excited state is 1. Thus, at some point, $\Omega(E, N, V)$ becomes a decreasing function of E, which implies a negative temperature. How could this be?

Exercise 3.5* Answer this question.

Assuming $(\partial \Omega / \partial E)_{N,V}$ is positive, the statistical postulate that at fixed N, V, and E all microstates are equally likely provides a molecular foundation for the theory of thermodynamics. The many results derived during our discussion of that topic (concerning stability, phase equilibria, Maxwell relations, etc.) are all consequences of this single fundamental law of nature.

3.3 Canonical Ensemble

When applying the microcanonical ensemble, the natural variables characterizing the macroscopic state of the system are E, V, and N. As we have seen in the context of thermodynamics, it is often convenient to employ other variables, and various representations of thermodynamics are found by applying Legendre transforms. In statistical mechanics, these manipulations are related to changes in ensembles. As an important example, we consider now the *canonical ensemble*—the assembly of all microstates with fixed N and V. The energy can fluctuate, however, and the system is kept at equilibrium by being in contact with a heat bath at temperature T (or inverse temperature β).

Schematically, we might picture the ensemble as in Fig. 3.2. The states we refer to here with the label v are states of definite energy—eigenfunctions of Schrödinger's equation, $\mathcal{H}\psi_v = E_v\psi_v$.

A system for which the canonical ensemble is appropriate can be viewed as a subsystem of one for which the microcanonical is applicable. See Fig. 3.3. This observation allows us to derive the distribution law for states in the canonical ensemble.

To begin, consider the case where the bath is so large that the energy of the bath, E_B, is overwhelmingly larger than the energy of the system, E_v. Further, the bath is so large that the energy levels of the bath are a continuum and $d\Omega/dE$ is well defined. The energy in the system fluctuates because the system is in contact with the bath, but

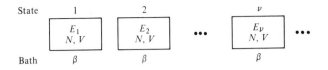

Fig. 3.2. Assembly of states for a closed system in a heat bath.

the sum $E = E_B + E_\nu$ is a constant. If the system is in one definite state ν, the number of states accessible to the system plus bath is $\Omega(E_B) = \Omega(E - E_\nu)$. Therefore, according to the statistical postulate—the principle of equal weights—the equilibrium probability for observing the system in state ν obeys

$$P_\nu \propto \Omega(E - E_\nu) = \exp\left[\ln \Omega(E - E_\nu)\right].$$

Since $E_\nu \ll E$, we can expand $\ln \Omega(E - E_\nu)$ in the Taylor series

$$\ln \Omega(E - E_\nu) = \ln \Omega(E) - E_\nu(d \ln \Omega/dE) + \cdots.$$

We choose to expand $\ln \Omega(E)$ rather than $\Omega(E)$ itself because the latter is a much more rapidly varying function of E than the former. We believe this because the formula $S = k_B \ln \Omega$ suggests that $\ln \Omega$ is relatively well behaved.

By retaining only those terms exhibited explicitly in the expansion (which is valid because the bath is considered to be an infinite thermal reservoir), and noting $(\partial \ln \Omega/\partial E)_{N,V} = \beta$, we obtain

$$P_\nu \propto \exp(-\beta E_\nu),$$

which is the canonical (or Boltzmann) distribution law. The constant of proportionality is independent of the specific state of the system and is determined by the normalization requirement

$$\sum_\nu P_\nu = 1.$$

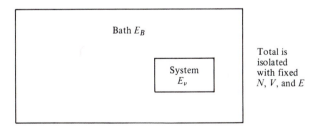

Fig. 3.3. A canonical ensemble system as a subsystem to microcanonical subsystem.

Hence,

$$P_v = Q^{-1} \exp(-\beta E_v),$$

where

$$Q(\beta, N, V) = \sum_v e^{-\beta E_v}.$$

The function $Q(\beta, N, V)$ is called the *canonical partition function*. It depends upon N and V through the dependence of the E_v's on these variables.

As an instructive example of its use, consider the calculation of the internal energy $E(\beta, N, V)$, which is $\langle E \rangle$ in the canonical distribution:

$$\langle E \rangle = \langle E_v \rangle = \sum_v P_v E_v$$

$$= \sum_v E_v e^{-\beta E_v} \Big/ \sum_{v'} e^{-\beta E_{v'}}$$

$$= -(\partial Q / \partial \beta)_{N,V} / Q$$

$$= -(\partial \ln Q / \partial \beta)_{N,V},$$

which suggests that $\ln Q$ is a familiar thermodynamic function. In fact, we will soon show that $-\beta^{-1} \ln Q$ is the Helmholtz free energy. For the next few pages, however, let us take this fact as simply given.

Exercise 3.6 Show that $(\partial \beta A / \partial \beta)_{N,V} = E$, where $A = E - TS$ is the Helmholtz free energy.

The energies E_v refer to the eigenvalues of Schrödinger's equation for the system of interest. In general, these energies are difficult, if not impossible, to obtain. It is significant, therefore, that a canonical ensemble calculation can be carried out independent of the exact solutions to Schrödinger's equation. This fact is understood as follows:

$$Q = \sum_v e^{-\beta E_v} = \sum_v \langle v | e^{-\beta \mathcal{H}} | v \rangle$$

$$= \text{Tr } e^{-\beta \mathcal{H}},$$

where "Tr" denotes the trace of a matrix (in this case, the trace of the Boltzmann operator matrix). It is a remarkable property of traces that they are independent of the representation of a matrix. (Proof: $\text{Tr } A = \text{Tr } SS^{-1}A = \text{Tr } S^{-1}AS$.) Thus, once we know $\mathcal{H}$ we can use *any* complete set of wavefunctions to compute Q. In other words, one may calculate $Q = \exp(-\beta A)$ without actually solving Schrödinger's equation with the Hamiltonian $\mathcal{H}$.

Exercise 3.7 Show that the internal energy, which is the average of E_v, can be expressed as $\mathrm{Tr}\,(\mathcal{H}e^{-\beta\mathcal{H}})/\mathrm{Tr}(e^{-\beta\mathcal{H}})$.

When calculating properties like the internal energy from the canonical ensemble, we expect that values so obtained should be the same as those found from the microcanonical ensemble. Indeed, as the derivation given above indicates, the two ensembles will be equivalent when the system is large. This point can be illustrated in two ways. First, imagine partitioning the sum over states in Q into groups of states with the same energy levels, that is

$$Q = \sum_{v(\text{states})} e^{-\beta E_v}$$

$$= \sum_{l(\text{levels})} \Omega(E_l)e^{-\beta E_l},$$

where we have noted that the number of states, $\Omega(E_l)$, is the degeneracy of the lth energy level. For a very large system, the spacing between levels is very small, and it seems most natural to pass to the continuum limit

$$Q \to \int_0^\infty dE\,\bar{\Omega}(E)e^{-\beta E},$$

where $\bar{\Omega}(E)$ is the density of states. In other words, for large systems, the canonical partition function is the *Laplace transform* of the microcanonical $\bar{\Omega}(E)$. An important theorem of mathematics is that Laplace transforms are unique. Due to this uniqueness, the two functions contain the identical information.

Nevertheless, energy fluctuates in the canonical ensemble while energy is fixed in the microcanonical ensemble. This inherent difference between the two does not contradict the equivalency of ensembles, however, because the relative size of the fluctuations becomes vanishingly small in the limit of large systems. To see why, let us compute the averaged square fluctuation in the canonical ensemble:

$$\langle(\delta E)^2\rangle = \langle(E - \langle E\rangle)^2\rangle$$

$$= \langle E^2\rangle - \langle E\rangle^2$$

$$= \sum_v P_v E_v^2 - \left(\sum_v P_v E_v\right)^2$$

$$= Q^{-1}(\partial^2 Q/\partial\beta^2)_{N,V} - Q^{-2}(\partial Q/\partial\beta)^2_{N,V}$$

$$= (\partial^2 \ln Q/\partial\beta^2)_{N,V}$$

$$= -(\partial\langle E\rangle/\partial\beta)_{N,V}.$$

Noting the definition of heat capacity, $C_v = (\partial E/\partial T)_{N,V}$, we have

$$\langle (\delta E)^2 \rangle = k_B T^2 C_v,$$

which is a remarkable result in its own right since it relates the size of spontaneous fluctuations, $\langle (\delta E)^2 \rangle$, to the rate at which energy will change due to alterations in the temperature. (The result foreshadows the topics of linear response theory and the fluctuation-dissipation theorem, which we will discuss in Chapter 8.) In the present context, we use the fluctuation formula to estimate the relative r.m.s. value of the fluctuations. Since the heat capacity is extensive, it is of order N (where N is the number of particles in the system). Furthermore $\langle E \rangle$ is also of order N. Hence the ratio of the dispersion to the average value is of order $N^{-1/2}$; that is,

$$\frac{\sqrt{\langle [E - \langle E \rangle]^2 \rangle}}{\langle E \rangle} = \frac{\sqrt{k_B T^2 C_v}}{\langle E \rangle} \sim O\left(\frac{1}{\sqrt{N}}\right).$$

For a large system ($N \sim 10^{23}$) this is a very small number and we may thus regard the average value, $\langle E \rangle$, as a meaningful prediction of the experimental internal energy. (For an ideal gas of structureless particles, $C_v = \frac{3}{2}Nk_B$, $\langle E \rangle = \frac{3}{2}Nk_B T$. Suppose $N \sim 10^{22}$, then the ratio above is numerically $\sim 10^{-11}$.) Furthermore, the microcanonical E, when written as a function of β, N, V by inverting $(\partial \ln \Omega/\partial E)_{N,V} = \beta(E, N, V)$, will be indistinguishable from the canonical internal energy $\langle E \rangle$ provided the system is large.

Exercise 3.8 Note that the probability for observing a closed thermally equilibrated system with a given energy E is $P(E) \propto \Omega(E)e^{-\beta E} = \exp[\ln \Omega(E) - \beta E]$. Both $\ln \Omega(E)$ and $-\beta E$ are of the order of N, which suggests that $P(E)$ is a very narrow distribution centered on the most probable value of E. Verify this suggestion by performing a steepest descent calculation with $P(E)$. That is, expand $\ln P(E)$ in powers of $\delta E = E - \langle E \rangle$, and truncate the expansion after the quadratic term. Use this expansion to estimate for 0.001 moles of gas the probability for observing a spontaneous fluctuation in E of the size of $10^{-6}\langle E \rangle$.

3.4 A Simple Example

To illustrate the theory we have been describing, consider a system of N distinguishable independent particles each of which can exist in one of two states separated by an energy ε. We can specify the state of a system, v, by listing

$$v = (n_1, n_2, \ldots, n_j, \ldots, n_N), \qquad n_j = 0 \text{ or } 1,$$

where n_j gives the state of particle j. The energy of the system for a given state is

$$E_v = \sum_{j=1}^{N} n_j \varepsilon,$$

where we have chosen the ground state as the zero of energy.

To calculate the thermodynamic properties of this model, we first apply the microcanonical ensemble. The degeneracy of the mth energy level is the number of ways one may choose m objects from a total of N. That is,

$$\Omega(E, N) = N!/(N - m)! \, m!,$$

where

$$m = E/\varepsilon.$$

The entropy and temperature are given by

$$S/k_B = \ln \Omega(E, N)$$

and

$$\beta = 1/k_B T = (\partial \ln \Omega/\partial E)_N$$
$$= \varepsilon^{-1}(\partial \ln \Omega/\partial m)_N.$$

For the last equality to have meaning, N must be large enough that $\Omega(E, N)$ will depend upon m in a continuous way. The continuum limit of factorials is *Stirling's approximation*: $\ln M! \approx M \ln M - M$, which becomes exact in the limit of large M. With that approximation

$$\frac{\partial}{\partial m} \ln \frac{N!}{(N - m)! m!} = -\frac{\partial}{\partial m} [(N - m) \ln (N - m)$$
$$- (N - m) + m \ln m - m]$$
$$= \ln \left(\frac{N}{m} - 1\right).$$

Combining this result with the formula for β yields

$$\beta\varepsilon = \ln \left(\frac{N}{m} - 1\right)$$

or

$$\frac{m}{N} = \frac{1}{1 + e^{\beta\varepsilon}}.$$

As a result, the energy $E = m\varepsilon$ as a function of temperature is

$$E = N\varepsilon \frac{1}{1 + e^{\beta\varepsilon}} \, ,$$

which is 0 at $T = 0$ (i.e., only the ground state is populated), and it is $N\varepsilon/2$ as $T \to \infty$ (i.e., all states are then equally likely).

Exercise 3.9 Use Stirling's approximation together with the formula for m/N to derive an expression for $S(\beta, N)$. Show that as $\beta \to \infty$ (i.e., $T \to 0$), S tends to zero. Find $S(E, N)$ and examine the behavior of $1/T$ as a function of E/N. Show that for some values of E/N, $1/T$ can be negative.

Of course, we could also study this model system with the canonical ensemble. In that case, the link to thermodynamics is

$$-\beta A = \ln Q = \ln \sum_{v} e^{-\beta E_v}.$$

Use of the formula for E_v gives

$$Q(\beta, N) = \sum_{n_1, n_2, \ldots n_N = 0, 1} \exp\left[-\beta \sum_{j=1}^{N} \varepsilon n_j\right]$$

since the exponential factors into an uncoupled product,

$$Q(\beta, N) = \prod_{j=1}^{N} \sum_{n_j = 0, 1} e^{-\beta \varepsilon n_j}$$

$$= (1 + e^{-\beta\varepsilon})^N.$$

As a result

$$-\beta A = N \ln (1 + e^{-\beta\varepsilon}).$$

The internal energy is

$$\langle E \rangle = \left(\frac{\partial(-\beta A)}{\partial(-\beta)}\right)_N = N\frac{\varepsilon e^{-\beta\varepsilon}}{1 + e^{-\beta\varepsilon}}$$

$$= N\varepsilon(1 + e^{\beta\varepsilon})^{-1},$$

in precise agreement with the result obtained with the microcanonical ensemble.

Exercise 3.10 Determine the entropy using

$$-\beta(A - \langle E \rangle) = S/k_B$$

and show the result is the same as that obtained for large N from the microcanonical ensemble.

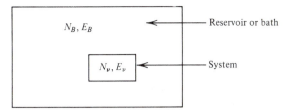

Fig. 3.4. System immersed in a bath.

3.5 Generalized Ensembles and the Gibbs Entropy Formula

Let us consider now, in a rather general way, why changes in ensembles correspond thermodynamically to performing Legendre transforms of the entropy. To begin, consider a system with X denoting the mechanical extensive variables. That is, $S = k_B \ln \Omega(E, X)$, and

$$k_B^{-1} \, dS = \beta \, dE + \xi \, dX.$$

For example, if $X = N$, then $\xi = -\beta\mu$. Or if X was the set of variables $V, N_1, N_2, \ldots$, then ξ would correspond to the conjugate set $\beta p, -\beta\mu_1, -\beta\mu_2, \ldots$, respectively. The quantity $-\xi/\beta$ therefore corresponds to f of Chapters 1 and 2.

Imagine an equilibrated system in which E and X can fluctuate. It can be viewed as a part of an isolated composite system in which the other part is a huge reservoir for E and X. An example could be an open system in contact with a bath with particles and energy flowing between the two. This example is pictured in Fig. 3.4.

The probability for microstates in the system can be derived in the same way we established the canonical distribution law. The result is

$$P_\nu = \exp(-\beta E_\nu - \xi X_\nu)/\Xi,$$

with

$$\Xi = \sum_\nu \exp(-\beta E_\nu - \xi X_\nu).$$

Exercise 3.11 Verify this result.

The thermodynamic E and X are given by the averages

$$\langle E \rangle = \sum_\nu P_\nu E_\nu = \left[\frac{\partial \ln \Xi}{\partial(-\beta)} \right]_{\xi, Y}$$

and

$$\langle X \rangle = \sum_v P_v X_v = \left[\frac{\partial \ln \Xi}{\partial (-\xi)} \right]_{\beta, Y},$$

where Y refers to all the extensive variables that are not fluctuating in the system. In view of the derivative relationships,

$$d \ln \Xi = - \langle E \rangle \, d\beta - \langle X \rangle \, d\xi.$$

Now consider the quantity

$$\mathscr{S} = -k_B \sum_v P_v \ln P_v.$$

We have

$$\mathscr{S} = -k_B \sum_v P_v [-\ln \Xi - \beta E_v - \xi X_v]$$

$$= k_B \{ \ln \Xi + \beta \langle E \rangle + \xi \langle X \rangle \}.$$

Therefore, $\mathscr{S}/k_B$ is the Legendre transform that converts $\ln \Xi$ to a function of $\langle E \rangle$ and $\langle X \rangle$; that is,

$$d\mathscr{S} = \beta k_B d \langle E \rangle + \xi k_B d \langle X \rangle,$$

which implies the $\mathscr{S}$ is, in fact, the entropy S. Thus in general

$$\boxed{S = -k_B \sum_v P_v \ln P_v.}$$

This result for the entropy is a famous one. It is called the Gibbs entropy formula.

Exercise 3.12 Verify that the microcanonical $S = k_B \ln \Omega(N, V, E)$ is consistent with the Gibbs formula.

The most important example of these formulas is that of the *grand canonical ensemble*. This ensemble is the assembly of all states appropriate to an *open* system of volume V. Both energy and particle number can fluctuate from state to state, and the conjugate fields that control the size of these fluctuations are β and $-\beta\mu$, respectively. Thus, letting v denote the state with N_v particles and energy E_v, we have

$$P_v = \Xi^{-1} \exp(-\beta E_v + \beta \mu N_v),$$

and the Gibbs entropy formula yields

$$S = -k_B \sum_v P_v[-\ln \Xi - \beta E_v + \beta \mu N_v]$$
$$= -k_B[-\ln \Xi - \beta \langle E \rangle + \beta \mu \langle N \rangle],$$

or, on rearranging terms

$$\ln \Xi = \beta p V,$$

where p is the thermodynamic pressure. Note that

$$\Xi = \sum_v \exp(-\beta E_v + \beta \mu N_v)$$

is a function of β, $\beta \mu$, and the volume. (It depends upon volume because the energies E_v depend upon the size of the system.) Hence, the "free energy" for an open system, $\beta p V$, is a natural function of β, $\beta \mu$, and V.

Fluctuation formulas in the grand canonical ensemble are analyzed in the same fashion as in the canonical ensemble. For example,

$$\langle (\delta N)^2 \rangle = \langle (N - \langle N \rangle)^2 \rangle = \langle N^2 \rangle - \langle N \rangle^2$$
$$= \sum_v N_v^2 P_v - \sum_v \sum_{v'} N_v N_{v'} P_v P_{v'}$$
$$= [\partial^2 \ln \Xi / \partial (\beta \mu)^2]_{\beta,V},$$

or

$$\langle (\delta N)^2 \rangle = (\partial \langle N \rangle / \partial \beta \mu)_{\beta,V}.$$

Generalizations to multicomponent systems can also be worked out in the same way, and they are left for the Exercises.

Recall that in our study of thermodynamic stability (i.e., the convexity of free energies) we found that $(\partial n/\partial \mu) \geq 0$. Now we see the same result in a different context. In particular, note that $\langle N \rangle = n N_0$, where N_0 is Avogadro's number, and since $\delta N = N - \langle N \rangle$ is a real number its square is positive. Hence, $\partial \langle N \rangle / \partial \beta \mu = \langle (\delta N)^2 \rangle \geq 0$. Similarly, in Chapter 2, we found from thermodynamic stability that $C_v \geq 0$, and in this chapter we learn that $k_B T^2 C_v = \langle (\delta E)^2 \rangle \geq 0$. In general, statistical mechanics will always give

$$-(\partial \langle X \rangle / \partial \xi) = \langle (\delta X)^2 \rangle.$$

The right-hand side is manifestly positive, and the left-hand side determines the curvature or convexity of a thermodynamic free energy.

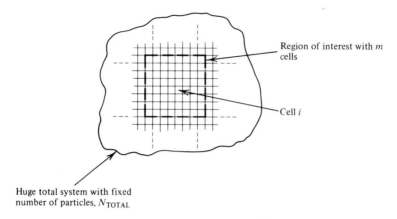

Fig. 3.5. Partitioning into cells.

3.6 Fluctuations Involving Uncorrelated Particles

In this section we will illustrate how the nature of spontaneous microscopic fluctuations governs the macroscopic observable behavior of a system. In the illustration, we consider concentration or density fluctuations in a system of uncorrelated particles, and we show that the ideal gas law (i.e., $pV = nRT$) follows from the assumption of no interparticle correlations. We will return to the ideal gas in Chapter 4 where we will derive its thermodynamic properties from detailed considerations of its energy levels. The following analysis, however, is of interest due to its generality being applicable even to large polymers at low concentration in a solvent.

To begin we imagine partitioning the volume of a system with cells as pictured in Fig. 3.5. Fluctuations in the region of interest follow the grand canonical distribution law described in Sec. 3.5. We will assume that the cells are constructed to be so small that there is a negligible chance for more than one particle to be in the same cell at the same time. Therefore, we can characterize any statistically likely configurations by listing the numbers $(n_1, n_2, \ldots, n_m)$, where

$$n_i = 1, \text{ if a particle is in cell } i$$
$$= 0, \text{ otherwise.}$$

In terms of these numbers, the instantaneous total number of particles in the region of interest is

$$N = \sum_{i=1}^{m} n_i,$$

and the mean square fluctuation in this number is

$$\langle (\delta N)^2 \rangle = \langle [N - \langle N \rangle]^2 \rangle = \langle N^2 \rangle - \langle N \rangle^2$$

$$= \sum_{i,j=1}^{m} [\langle n_i n_j \rangle - \langle n_i \rangle \langle n_j \rangle]. \tag{a}$$

These relations are completely general. A simplification is found by considering the case in which different particles are uncorrelated with each other and this lack of correlation is due to a very low concentration of particles. These two physical considerations imply

$$\langle n_i n_j \rangle = \langle n_i \rangle \langle n_j \rangle \quad \text{for} \quad i \neq j \tag{b}$$

(see Exercise 3.17), and

$$\langle n_i \rangle \ll 1, \tag{c}$$

respectively. Further, since n_i is either zero or one, $n_i^2 = n_i$ and hence

$$\langle n_i^2 \rangle = \langle n_i \rangle = \langle n_1 \rangle, \tag{d}$$

where the last equality follows from the assumption each cell is of the same size or type. Hence, on the average, each cell behaves identically to every other cell.

The insertion of (b) into (a) yields

$$\langle (\delta N)^2 \rangle = \sum_{i=1}^{m} [\langle n_i^2 \rangle - \langle n_i \rangle^2],$$

and the application of (d) gives

$$\langle (\delta N)^2 \rangle = m \langle n_1 \rangle (1 - \langle n_1 \rangle).$$

Finally, from (c) we arrive at

$$\langle (\delta N)^2 \rangle \approx m \langle n_1 \rangle = \langle N \rangle.$$

By itself, this relationship is already a remarkable result, but its thermodynamic ramification is even more impressive.

In particular, since the region of interest is described by the grand canonical ensemble, we know that (see Sec. 3.5 and Exercise 3.15)

$$\langle (\delta N)^2 \rangle = (\partial \langle N \rangle / \partial \beta \mu)_{\beta, V}.$$

Hence, for a system of uncorrelated particles, we have

$$(\partial \langle N \rangle / \partial \beta \mu)_{\beta, V} = \langle N \rangle,$$

or dividing by V and taking the reciprocal

$$(\partial \beta \mu / \partial \rho)_\beta = \rho^{-1},$$

where $\rho = \langle N \rangle / V$. Thus, by integration we have

$$\beta\mu = \text{constant} + \ln \rho.$$

Further, from standard manipulations (see Exercise 1.14)

$$(\partial \beta p / \partial \rho)_\beta = \rho(\partial \beta\mu / \partial \rho)_\beta = 1,$$

where the first equality is a general thermodynamic relation and the second applies what we have discovered for uncorrelated particles. Integration yields

$$\beta p = \rho,$$

where we have set the constant of integration to zero since the pressure should vanish as the density ρ goes to zero. This equation is the celebrated ideal gas law, $pV = nRT$, where we identify the gas constant, R, with Boltzmann's constant times Avogadro's number, N_0:

$$R = k_B N_0.$$

In summary, we have shown that the assumption of uncorrelated statistical behavior implies that for a one-component system

$$\rho \propto e^{\beta\mu}$$

and

$$\beta p / \rho = 1.$$

Generalizations to multicomponent systems are straightforward and left for Exercises.

3.7 Alternative Development of Equilibrium Distribution Functions

The approach we have followed thus far begins with a statistical characterization of equilibrium states and then arrives at the inequalities and distribution laws we regard as the foundation of thermodynamics. Alternatively, we could begin with the second law and the Gibbs entropy formula rather than deducing them from the principle of equal weights. In the next few pages we follow this alternative development.

Extensivity of Entropy

Since the Gibbs entropy formula is our starting point, let's check that it satisfies the additivity property (extensivity) that we associate with

Fig. 3.6. Two independent subsystems A and B.

entropy. Consider a system contained in two boxes, A and B (see Fig. 3.6). Denote the total entropy of the system by S_{AB}. If the entropy is extensive, $S_{AB} = S_A + S_B$. From the Gibbs formula

$$S_{AB} = -k_B \sum_{v_A} \sum_{v_B} P_{AB}(v_A, v_B) \ln P_{AB}(v_A, v_B),$$

where v_A and v_B denote states of the subsystems A and B, respectively. Since the subsystems are uncoupled,

$$P_{AB}(v_A, v_B) = P_A(v_A) P_B(v_B).$$

Thus

$$S_{AB} = -k_B \sum_{v_A} \sum_{v_B} P_{AB}(v_A, v_B)[\ln P_{AB}(v_A, v_B)]$$

$$= -k_B \sum_{v_B} P_B(v_B) \sum_{v_A} P_A(v_A) \ln P_A(v_A)$$

$$-k_B \sum_{v_A} P_A(v_A) \sum_{v_B} P_B(v_B) \ln P_B(v_B)$$

$$= -k_B \sum_{v_A} P_A(v_A) \ln P_A(v_A) - k_B \sum_{v_B} P_B(v_B) \ln P_B(v_B)$$

$$= S_A + S_B,$$

where the second to last equality is obtained from the normalization condition. This simple calculation shows that the Gibbs entropy exhibits the thermodynamic property $S_{AB} = S_A + S_B$.

Exercise 3.13 Show that if one assumes the functional form

$$S = \sum_v P_v f(P_v),$$

where $f(x)$ is some function of x, then the requirement that S is extensive implies that $f(x) = c \ln x$, where c is an arbitrary constant.

The Microcanonical Ensemble

For an isolated system, the energy, E, the number of particles, N, and the volume, V, are fixed. The ensemble appropriate to such a system is the *microcanonical ensemble*: the assembly of all states with E, N, and V fixed.

To derive the equilibrium probability for state j, P_j, we require that the condition for thermodynamic equilibrium be satisfied. According to the second law,

$$(\delta S)_{E,V,N} = 0.$$

In other words, the partitioning of microscopic states at equilibrium is the partitioning that maximizes the entropy. We use this principle and carry out a maximization procedure with the constraints

$$\langle E \rangle = \sum_j E_j P_j, \tag{a}$$

$$\langle N \rangle = \sum_j N_j P_j, \tag{b}$$

and

$$1 = \sum_j P_j. \tag{c}$$

In the microcanonical ensemble where $E_j = E = \text{constant}$, and $N_j = N = \text{constant}$, conditions (a), (b), and (c) are all the same.

Using the Lagrange multiplier γ, we seek a P_j for which

$$\delta(S + \gamma 1) = 0,$$

or, inserting Eq. (c) and the Gibbs entropy formula,

$$0 = \delta \left\{ -k_B \sum_j P_j \ln P_j + \gamma \sum_j P_j \right\}$$

$$= \sum_j \delta P_j [-k_B \ln P_j - k_B + \gamma].$$

For this equation to be valid for all δP_j, the quantity in [ ] must equal zero. Thus

$$\ln P_j = \frac{\gamma - k_B}{k_B} = \text{constant}.$$

The constant can be determined from the normalization condition

$$1 = \sum_j P_j = \sum_j e^{\text{constant}} \equiv \sum_j \frac{1}{\Omega} = \frac{1}{\Omega} \left(\sum_j 1 \right).$$

Thus

$$\Omega = \text{the number of states with energy } E.$$

In summary, for the microcanonical ensemble

$$P_j = \frac{1}{\Omega}, \quad \text{for } E_j = E$$

$$= 0, \quad \text{for } E_j \neq E$$

and the entropy is

$$S = +k_B \sum_j \frac{1}{\Omega} \ln \Omega = k_B \ln \Omega \sum_j \frac{1}{\Omega} = k_B \ln \Omega.$$

The Canonical Ensemble

This ensemble is appropriate to a closed system in a temperature bath. N, V, and T are fixed, but the energy is not. Thermodynamic equilibrium now gives

$$\delta(S + \alpha \langle E \rangle + \gamma 1) = 0,$$

where α and γ are Lagrange multipliers. By combining Eqs. (a), (c), and the Gibbs entropy formula with the above, we obtain

$$\sum_j [-k_B \ln P_j - k_B + \alpha E_j + \gamma] \, \delta P_j = 0.$$

For this expression to hold for all δP_j,

$$[-k_B \ln P_j - k_B + \alpha E_j + \gamma] = 0$$

or

$$\ln P_j = \frac{\alpha E_j - k_B + \gamma}{k_B}. \tag{d}$$

To determine α and γ, we use the thermodynamic identity

$$\left[\frac{\delta \langle E \rangle}{\delta S} \right]_{V,N} = T = \text{temperature}.$$

With Eq. (a) we find

$$(\delta \langle E \rangle)_{V,N} = \sum_j E_j \, \delta P_j,$$

and from the Gibbs entropy formula and (d) we find

$$(\delta S)_{V,N} = -k_B \sum_j \delta P_j \left[\frac{\alpha E_j - k_B + \gamma}{k_B} \right]$$

$$= -k_B \sum_j \delta P_j E_j \alpha / k_B,$$

where the last equality follows from the fact that $\sum_j \delta P_j = \delta 1 = 0$. Note that in the variation of $\langle E \rangle$, we do not alter E_j since the variation refers to changes in P_j (i.e., partitioning of states) with the energies of the states fixed. Dividing $(\delta \langle E \rangle)_{V,N}$ by $(\delta S)_{V,N}$ yields

$$T = \left[\frac{\delta \langle E \rangle}{\delta S} \right]_{V,N} = -\frac{1}{\alpha}.$$

Combining this result with (d) and the Gibbs entropy formula gives

$$S = \sum_j P_j \left[\frac{E_j + k_B T - \gamma T}{T} \right]$$
$$= \frac{\langle E \rangle + k_B T - \gamma T}{T}.$$

Thus

$$\gamma T = A + k_B T,$$

where

$$A = \langle E \rangle - TS = \text{Helmholtz free energy}.$$

In summary, the canonical ensemble has

$$P_j = e^{-\beta[E_j - A]},$$

where

$$\beta = \frac{1}{k_B T}.$$

Since P_j is normalized,

$$\sum_j P_j = 1 = e^{\beta A} \sum_j e^{-\beta E_j}.$$

Thus, the *partition function, Q,*

$$Q = \sum_j e^{-\beta E_j}$$

is also given by

$$Q = e^{-\beta A}.$$

From thermodynamic considerations alone, it is clear that the knowledge of Q tells us everything about the thermodynamics of our system. For example,

$$\left[\frac{\partial \ln Q}{\partial V} \right]_{T,N} = \left[\frac{\partial(-\beta A)}{\partial V} \right]_{T,N} = \beta p,$$

where p is the pressure, and

$$\left[\frac{\partial \ln Q}{\partial \beta}\right]_{V,N} = \left[\frac{\partial(-\beta A)}{\partial \beta}\right]_{V,N} = -\langle E \rangle,$$

where $\langle E \rangle$ is the internal energy.

Similar analysis can be applied to other ensembles, too. In general, therefore, the principle of equal weights is equivalent to the Gibbs entropy formula and the variational statement of the second law of thermodynamics.

Additional Exercises

3.14. By applying Gibbs entropy formula and the equilibrium condition

$$(\delta S)_{\langle E \rangle, V, \langle N \rangle} = 0,$$

derive the probability distribution for the grand canonical ensemble—the ensemble in which N and E can vary. Your result should be

$$P_\nu = \Xi^{-1} \exp\left[-\beta E_\nu + \beta \mu N_\nu\right],$$

where ν labels that state of the system (including the number of particles) and

$$\Xi = \exp(\beta p V).$$

3.15. For an open multicomponent system, show that

$$\langle \delta N_i \delta N_j \rangle = (\partial \langle N_i \rangle / \partial \beta \mu_j)_{\beta, \beta \mu_l, V},$$

where $\delta N_i = N_i - \langle N_i \rangle$ is the fluctuation from the average of the number of particles of type i, and μ_i is the chemical potential for that type of particle. Similarly, relate $\langle \delta N_i \delta N_l \delta N_j \rangle$ to a thermodynamic derivative. Finally, for a one-component system in the grand canonical ensemble, evaluate $\langle (\delta E)^2 \rangle$ and relate this quantity to the constant volume heat capacity and the compressibility. The former determines the size of the mean square energy fluctuations in the canonical ensemble where density does not fluctuate, and the latter determines the size of the mean square density fluctuations.

3.16. For 0.01 moles of ideal gas in an open thermally equilibrated system, evaluate numerically the relative root mean square deviation of the energy from its mean value and the relative root mean square deviation of the density from its mean value.

3.17. (a) Consider a random variable x that can take on any value in the interval $a \leq x \leq b$. Let $g(x)$ and $f(x)$ be any functions of x and let $\langle \cdots \rangle$ denote the average over the distribution for x, $p(x)$—that is,

$$\langle g \rangle = \int_a^b dx \, g(x) p(x).$$

Show that

$$\langle gf \rangle = \langle g \rangle \langle f \rangle$$

for arbitrary $g(x)$ and $f(x)$ if and only if

$$p(x) = \delta(x - x_0),$$

where x_0 is a point between a and b, and $\delta(x - x_0)$ is the Dirac delta function,

$$\delta(y) = 0, \qquad y \neq 0$$

and

$$\int_{-\varepsilon}^{\varepsilon} dy \, \delta(y) = 1.$$

Note that according to this definition, $\delta(x - x_0)$ is a normalized distribution of zero (or infinitesimal) width located at $x = x_0$.

(b) Consider two random variables x and y with the joint probability distribution $p(x, y)$. Prove that

$$\langle f(x) g(y) \rangle = \langle f \rangle \langle g \rangle$$

for all functions $f(x)$ and $g(y)$, if and only if

$$p(x, y) = p_1(x) p_2(y),$$

where $p_1(x)$ and $p_2(y)$ are the distributions for x and y, respectively.

3.18. Consider a system of N distinguishable non-interacting spins in a magnetic field H. Each spin has a magnetic moment of size μ, and each can point either parallel or antiparallel to the field. Thus, the energy of a particular state is

$$\sum_{i=1}^{N} -n_i \mu H, \qquad n_i = \pm 1,$$

where $n_i \mu$ is the magnetic moment in the direction of the field.

(a) Determine the internal energy of this system as a function of β, H, and N by employing an ensemble characterized by these variables.

(b) Determine the entropy of this system as a function of β, H, and N.

(c) Determine the behavior of the energy and entropy for this system as $T \to 0$.

3.19. (a) For the system described in Exercise 3.18, derive the average total magnetization,

$$\langle M \rangle = \left\langle \sum_{i=1}^{N} \mu n_i \right\rangle,$$

as a function of β, H, and N.

(b) Similarly, determine $\langle (\delta M)^2 \rangle$, where

$$\delta M = M - \langle M \rangle,$$

and compare your result with the susceptibility

$$(\partial \langle M \rangle / \partial H)_{\beta, N}.$$

(c) Derive the behavior of $\langle M \rangle$ and $\langle (\delta M)^2 \rangle$ in the limit $T \to 0$.

3.20. Consider the system studied in Exercises 3.18 and 3.19. Use an ensemble in which the total magnetization is fixed, and determine the magnetic field over temperature, βH, as a function of the natural variables for that ensemble. Show that in the limit of large N, the result obtained in this way is equivalent to that obtained in Exercise 3.19.

3.21.* In this problem you consider the behavior of mixed valence compounds solvated at very low concentrations in a crystal. Figure 3.7 is a schematic depiction of such a compound. We shall assume the compound has only two configurational states as illustrated in Fig. 3.8. The two states correspond to having the electron localized on the left or right iron atoms, respectively. This type of two-state model is like the LCAO

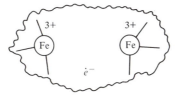

Fig. 3.7. A mixed valence compound conceived of as two cations plus an electron.

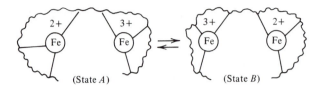

Fig. 3.8. Two-state model of a mixed valence compound.

treatment of the H_2^+ molecule in elementary quantum chemistry. In the solid state physics literature, the model is called the "tight binding" approximation.

In the absence of the surrounding crystal, the Hamiltonian for a compound is $\mathcal{H}_0$ with matrix elements

$$\langle A| \mathcal{H}_0 |A\rangle = \langle B| \mathcal{H}_0 |B\rangle = 0 \text{ (our choice for the}$$
$$\text{zero of energy),}$$

$$\langle A| \mathcal{H}_0 |B\rangle = -\Delta.$$

The dipole moment of one of the compounds for states A or B is given by

$$\mu = \langle A| m |A\rangle = -\langle B| m |B\rangle,$$

where m denotes the electronic dipole operator. For further simplicity, imagine that there is negligible spatial overlap between states A and B; that is,

$$\langle A \mid B\rangle = 0 \quad \text{and} \quad \langle A| m|B\rangle = 0.$$

The solvent crystal couples to the impurity mixed valence compounds through the electric crystal field, $\mathcal{E}$. The Hamiltonian for each compound is

$$\mathcal{H} = \mathcal{H}_0 - m\mathcal{E}.$$

(a) Show that when $\mathcal{E} = 0$, the eigenstates of the single compound Hamiltonian are

$$|\pm\rangle = \frac{1}{\sqrt{2}}[|A\rangle \pm |B\rangle],$$

and the energy levels are $\pm\Delta$.
(b) Compute the canonical partition function for the system of mixed valence compounds when $\mathcal{E} = 0$, by (i) performing the Boltzmann weighted sum with energy eigenvalues, and (ii) performing the matrix trace of $e^{-\beta\mathcal{H}_0}$ employing the configurational states $|A\rangle$ and $|B\rangle$. The latter states

diagonalize m but not $\mathcal{H}_0$. Nevertheless, the two computations should yield the same result. Why?

(c) When $\mathscr{E}$ is zero, determine the averages (i) $\langle m \rangle$, (ii) $\langle |m| \rangle$, and (iii) $\langle (\delta m)^2 \rangle$, where $\delta m = m - \langle m \rangle$.

(d) When $\mathscr{E} \neq 0$, the crystal couples to the impurity compounds, and there is a free energy of solvation, $[A(\mathscr{E}) - A(0)]/N$, where N is the number of compounds. Compute this free energy of solvation by (i) first determining the eigen energies as a function of $\mathscr{E}$, then performing the appropriate Boltzmann weighted sum, and (ii) performing the appropriate matrix trace employing configuration states $|A\rangle$ and $|B\rangle$. The two calculations yield the same result, though the second is algebraically more tedious. (You might find it useful to organize the algebra in the second case by exploring the properties of Pauli spin matrices.)

(e) When $\mathscr{E} \neq 0$, compute $\langle |m| \rangle$ and $\langle m \rangle$. Compare its value with what is found when $\mathscr{E} = 0$. Why does $\langle m \rangle$ increase with increasing $\mathscr{E}$?

3.22. (a) Consider a region within a fluid described by the van der Waals equation $\beta p = \rho/(1 - b\rho) - \beta a \rho^2$, where $\rho = \langle N \rangle / V$. The volume of the region is L^3. Due to the spontaneous fluctuations in the system, the instantaneous value of the density in that region can differ from its average by an amount $\delta \rho$. Determine, as a function of β, ρ, a, b, and L^3, the typical relative size of these fluctuations; that is, evaluate $\langle (\delta \rho)^2 \rangle^{1/2} / \rho$. Demonstrate that when one considers observations of a macroscopic system (i.e., the size of the region becomes macroscopic, $L^3 \to \infty$) the relative fluctuations become negligible.

(b) A fluid is at its "critical point" when

$$(\partial \beta p / \partial \rho)_\beta = (\partial^2 \beta p / \partial \rho^2)_\beta = 0.$$

Determine the critical point density and temperature for the fluid obeying the van der Waals equation. That is, compute β_c and ρ_c as a function of a and b.

(c) Focus attention on a subvolume of size L^3 in the fluid. Suppose L^3 is 100 times the space filling volume of a molecule—that is, $L^3 \approx 100b$. For this region in the fluid, compute the relative size of the density fluctuations when $\rho = \rho_c$, and the temperature is 10% above the critical temperature. Repeat this calculation for temperatures 0.1% and 0.001% from the critical temperature.

(d) Light that we can observe with our eyes has wavelengths of the order of 1000 Å. Fluctuations in density cause changes in the index of refraction, and those changes produce scattering of light. Therefore, if a region of fluid 1000 Å across contains significant density fluctuations, we will visually observe these fluctuations. On the basis of the type of calculation performed in part (b), determine how close to the critical point a system must be before critical fluctuations become optically observable. The phenomenon of long wavelength density fluctuations in a fluid approaching the critical point is known as critical opalescence. (Note: You will need to estimate the size of b, and to do this you should note that the typical diameter of a small molecule is around 5 Å.)

3.23. Consider a solution containing a solute species at very low concentrations. The solute molecules undergo conformational transitions between two isomers, A and B. Let N_A and N_B denote the numbers of A and B isomers, respectively. While the total number of solute molecules $N = N_A + N_B$ remain constant, at any instant the values of N_A and N_B differ from their mean values of $\langle N_A \rangle$ and $\langle N_B \rangle$. Show that mean square fluctuations are given by

$$\langle (N_A - \langle N_A \rangle)^2 \rangle = x_A x_B N,$$

where x_A and x_B are the average mole fractions of A and B species; that is,

$$x_A = \langle N_A \rangle / N.$$

[*Hint:* You will need to assume that the solutes are at such a low concentration that each solute molecule is uncorrelated from every other solute molecule. See Sec. 3.6.]

Bibliography

There are many excellent texts treating the basic elementary principles of statistical mechanics. For example:

T. L. Hill, *Introduction to Statistical Thermodynamics* (Addison-Wesley, Reading, Mass., 1960).

G. S. Rushbrooke, *Introduction to Statistical Mechanics* (Oxford University Press, Oxford, 1951).

D. McQuarrie, *Statistical Mechanics* (Harper & Row, N.Y., 1976).

F. Reif, *Fundamentals of Statistical and Thermal Physics* (McGraw-Hill, N.Y., 1965).

The following are somewhat more advanced than this text:

S. K. Ma, *Statistical Mechanics* (World Scientific, Philadelphia, 1985).

R. Balescu, *Equilibrium and Nonequilibrium Statistical Mechanics* (John Wiley, N.Y., 1975).

M. Toda, R. Kubo, and N. Saito, *Statistical Physics I* (Springer-Verlag, N.Y., 1983).

Each contains chapters that introduce modern ideas about ergodicity and chaos.

A colorful discussion of the history and principles of statistical mechanics is given by

P. W. Atkins, *The Second Law* (Scientific American Books and
W. H. Freeman and Co., N.Y., 1984).

This book also contains instructive illustrations of concepts concerning chaos and chaotic structures including codes to run on microcomputers.

CHAPTER 4

Non-Interacting (Ideal) Systems

In this chapter we consider the simplest systems treated by statistical mechanics. These are systems composed of particles (or quasi-particles) that do not interact with each other; these models are called ideal gases.

The principles of statistical mechanics prescribe the computation of partititon functions like

$$\sum_v \exp\left(-\beta E_v\right)$$

or

$$\sum_v \exp\left[-\beta(E_v - \mu N_v)\right].$$

These are Boltzmann weighted sums over all possible fluctuations, that is, all microscopic states permitted by the constraints with which we control the system. In the first sum, only states with the same number of particles are considered; in the second, particle numbers fluctuate, too, and the chemical potential term accounts for the energetics of changing particle number. Notice that if we restricted the second sum to include only those states v for which the number of particles N_v had the value N, then the second sum would be proportional to the first.

These sums or partition functions are central to the theory since the probability for something to occur is the Boltzmann weighted sum over all fluctuations or microstates consistent with that occurrence. For instance, in an open system where the number of particles, N_v, can fluctuate from state to state, the probability of having precisely N particles is

$$P_N \propto \sum_{\nu}^{(N)} \exp\left[-\beta(E_\nu - \mu N_\nu)\right] = e^{\beta \mu N} \sum_{\nu_N} \exp\left(-\beta E_{\nu_N}\right),$$

where the superscript "N" on the summation indicates that the sum includes only those states ν for which $N_\nu = N$, and those states are labeled with the index ν_N.

In the absence of constraints, fluctuations occur spontaneously, and these formulas show that the likelihood of spontaneous fluctuations is governed by energetics of these fluctuations as compared with Boltzmann's thermal energy, $k_B T = \beta^{-1}$. Thus, higher T allows for greater fluctuations or randomness, and as $T \to 0$, only those states for which the energy per particle is the same as the energy per particle of the ground state are accessible.

The systematic exploration of all possible fluctuations is often a very complicated task due to the huge number of microscopic states that must be considered, and the cumbersome detail needed to characterize these states. This complexity is the reason why statistical mechanics is often regarded as a difficult subject. As we proceed in this book, however, we will introduce the reader to a number of practical methods for sampling relevant fluctuations. The simplest of these are factorization approximations that become exact when the system is composed of non-interacting degrees of freedom—the class of models considered in this chapter.

To understand how the factorization method works, suppose the energy E_ν breaks into two parts: $E_\nu = E_n^{(1)} + E_m^{(2)}$, where the state label ν depends upon n and m, and these indices m and n are independent of each other. Then the canonical partition function,

$$Q = \sum_\nu e^{-\beta E_\nu}$$

$$= \sum_{n,m} \exp\left(-\beta E_n^{(1)}\right) \exp\left(-\beta E_m^{(2)}\right),$$

can be factored as

$$Q = \left[\sum_n \exp\left(-\beta E_n^{(1)}\right)\right]\left[\sum_m \exp\left(-\beta E_m^{(2)}\right)\right]$$

$$= Q^{(1)} Q^{(2)},$$

where the second equality introduces $Q^{(1)}$ and $Q^{(2)}$ as the Boltzmann weighted sums associated with the energies $E_n^{(1)}$ and $E_m^{(2)}$, respectively. Notice that these energies are uncorrelated in the sense that

$$\langle E^{(1)}E^{(2)}\rangle = Q^{-1} \sum_{n,m} E_n^{(1)} E_m^{(2)} \exp\left[-\beta(E_n^{(1)} + E_m^{(2)})\right]$$
$$= [\partial \ln Q^{(1)}/\partial(-\beta)] [\partial \ln Q^{(2)}/\partial(-\beta)]$$
$$= \langle E^{(1)}\rangle\langle E^{(2)}\rangle.$$

Generalization to the case where there are N uncorrelated degrees of freedom is not difficult, and one finds

$$Q = Q^{(1)}Q^{(2)} \cdots Q^{(N)}.$$

If each of these degrees of freedom is of the same type, the formula further reduces to

$$Q = [Q^{(1)}]^N.$$

This factorization therefore implies that we only need to Boltzmann sample microstates for one degree of freedom and then simply take the Nth power of this result. To dramatize the significance of such a simplification, suppose a system had $N = 1000$ degrees of freedom, and each could exist in one of five microstates. The total number of states to be sampled for the entire system is 5^{1000}—an impossibly large number. The factorization, however, implies that we need only explicitly enumerate five states.

In some cases, the factorization approximation is applicable because the system is composed of uncorrelated particles. An example is a classical ideal gas. Here, the energy is a sum of one-particle energies, and, if the particles were distinguishable, the partition function would be simply q^N, where q is the Boltzmann sum over states for a single particle. Further, at the high temperatures for which "classical" models are good approximations to reality (made precise later), the number of single particle states available is very large compared to the number of particles. In this case, each N-particle state occurs $N!$ times, corresponding to the number of ways of assigning the N distinct one-particle states to the N indistinguishable particles. Hence, the correct partition function is

$$\frac{1}{N!} q^N.$$

Without the factor of $(N!)^{-1}$, we would be overcounting the distinguishable states.

In other cases, the factorization approximation is applicable even when the actual particles in the system are not uncorrelated. Here, one finds that it is possible to identify uncorrelated *collective variables*—variables that depend upon the coordinates or states of a

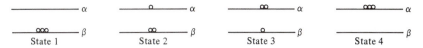

Fig. 4.1. The four states of a three-particle system with two single particle states.

large collection of particles. An example is the small amplitude vibrational modes of a solid. These modes are called *phonons*. Another example is the *occupation numbers* for quantum mechanical systems composed of non-interacting particles.

In this chapter, we will consider phonons, occupation numbers, classical ideal gases, and a number of other examples to illustrate how the factorization method is applied.

4.1 Occupation Numbers

The first step in analyzing any model involves the classification of microstates. The state of a quantum system can be specified by the wavefunction for that state, $\Psi_\nu(r_1, r_2, \ldots, r_N)$. Here, Ψ_ν is the νth eigensolution to Schrödinger's equation for an N-particle system. If the particles are non-interacting (i.e., ideal), then the wavefunction can be expressed as a symmetrized* product of single particle wavefunctions. Let us denote these single particle wavefunctions as $\phi_1(r)$, $\phi_2(r)$, $\ldots$, $\phi_j(r)$, $\ldots$. For a particular state, say ν, $\Psi_\nu(r_1, \ldots, r_N)$ will be a symmetrized product containing n_1 particles with the single particle wavefunction ϕ_1, n_2 particles with the single particle wavefunction ϕ_2, and so on. These numbers, $n_1, n_2, \ldots, n_j, \ldots$ are called the *occupation numbers* of the first, second, $\ldots$ jth, $\ldots$ single particle states. If the N particles are indistinguishable—as quantum particles are—then *a state, ν, is completely specified by the set of occupation numbers* $(n_1, n_2, \ldots, n_j, \ldots)$ since any more detail would distinguish between the n_j particles in the jth single particle state.

For example, consider three particles (denoted in Fig. 4.1 by circles) which can exist in one of two single particle states, α and β. All the possible states for this three-particle system are exhibited in Fig. 4.1. In terms of occupation numbers, state 1 has $n_\alpha = 0$, $n_\beta = 3$; state 2 has $n_\alpha = 1$, $n_\beta = 2$; and so on. Notice that an occupation number is a collective variable in the sense that its value depends upon the instantaneous state of all the particles.

Let us now express the total number of particles and the total

---

* For Fermi particles, the product is antisymmetric; for Bose particles the product is symmetric.

energy in terms of the occupation numbers. Let

$$v = (n_1, n_2, \ldots, n_j, \ldots) = v\text{th state.}$$

Then

$$N_v = \sum_j n_j = \text{total number of particles in the } v\text{th state.}$$

Let ε_j be the energy of the jth single particle state. Then,

$$E_v = \sum_j \varepsilon_j n_j = \text{energy in the } v\text{th state.}$$

Particles with half-integer spin obey an exclusion principle*: $n_j = 0$ or 1, only. Such particles are called *fermions* and the statistics associated with $n_j = 0$ or 1 is called *Fermi–Dirac* statistics.

Particles with integer spin obey Bose–Einstein statistics: $n_j = 0, 1, 2, 3, \ldots$. These particles are called *bosons*.

4.2 Photon Gas

As an example of how we use occupation numbers, consider the photon gas—an electromagnetic field in thermal equilibrium with its container. We want to describe the thermodynamics of this system. From the quantum theory of the electromagnetic field, it is found that the Hamiltonian can be written as a sum of terms, each having the form of a Hamiltonian for a harmonic oscillator of some frequency. The energy of a harmonic oscillator is $n\hbar\omega$ (zero point energy omitted), where $n = 0, 1, 2, \ldots$. Thus, we are led to the concept of photons with energy $\hbar\omega$. A state of the free electromagnetic field is specified by the number n for each of the "oscillators," and n can be thought of as the number of photons in a state with single "particle" energy $\hbar\omega$.

Photons obey Bose–Einstein statistics: $n = 0, 1, 2, \ldots$. The canonical partition function is thus

$$e^{-\beta A} = Q = \sum_v e^{-\beta E_v} = \sum_{\substack{n_1, n_2, \ldots, n_j, \ldots \\ =0}}^{\infty} e^{-\beta(n_1\varepsilon_1 + n_2\varepsilon_2 + \cdots + n_j\varepsilon_j + \cdots)},$$

where we have used the occupation number representation of E_v, and denoted $\hbar\omega_j$ by ε_j. Since the exponential factors into independ-

---

* The requirement that the N-particle wavefunction be an antisymmetric product implies the exclusion principle.

ent portions, we have

$$Q = \prod_j \left[\sum_{n_j=0}^{\infty} e^{-\beta n_j \varepsilon_j} \right].$$

The term in brackets is just a geometric series, thus

$$Q(\text{photon gas}) = \prod_j \left[\frac{1}{1 - e^{-\beta \varepsilon_j}} \right].$$

From this formula, we can obtain all the properties we want since $Q = e^{-\beta A}$. One quantity that is particularly interesting is the average value of the occupation number of the jth state, $\langle n_j \rangle$. In the canonical ensemble

$$\langle n_j \rangle = \frac{\sum\limits_{\nu} n_j e^{-\beta E_\nu}}{\sum\limits_{\nu} e^{-\beta E_\nu}} = \frac{\sum\limits_{n_1, n_2, \ldots} n_j e^{-\beta(n_1 \varepsilon_1 + \cdots + n_j \varepsilon_j + \cdots)}}{Q}$$

$$= \left[\frac{\partial}{\partial(-\beta \varepsilon_j)} \sum_{n_1, n_2, \ldots} e^{-\beta(n_1 \varepsilon_1 + \cdots + n_j \varepsilon_j + \cdots)} \right] \Big/ Q$$

$$= \frac{\partial \ln Q}{\partial(-\beta \varepsilon_j)}.$$

Returning to our formula for Q we thus have

$$\langle n_j \rangle = + \frac{\partial}{\partial(-\beta \varepsilon_j)} \left\{ \sum_j -\ln(1 - e^{-\beta \varepsilon_j}) \right\}$$

$$= e^{-\beta \varepsilon_j} / [1 - e^{-\beta \varepsilon_j}]$$

or

$$\langle n_j \rangle = [e^{\beta \varepsilon_j} - 1]^{-1},$$

which is called the Planck distribution.

Exercise 4.1 For the photon gas derive a formula for the *correlation function* $\langle \delta n_i \, \delta n_j \rangle$ where $\delta n_i = n_i - \langle n_i \rangle$.

Exercise 4.2* Use the formula for $\langle n_j \rangle$ to show that the energy density of a photon gas is σT^4, where σ is a constant, $(\pi^2 k_B^4 / 15 \hbar^3 c^3)$. [*Hint*: You will need to construct a formula for the number of standing wave solutions to use the wave equation for waves in a three-dimensional cavity and with frequency between ω and $\omega + d\omega$.]

4.3 Phonon Gas or Fluctuations of Atomic Positions in a Cold Solid

As another example, consider the phonon gas—the normal modes of a low temperature solid. For a cold enough lattice, the atoms remain close to their equilibrium positions. As a result, the potential energy of such a system can be expanded in powers of the displacement of coordinates from their equilibrium locations. For many applications, it is a good approximation to truncate this expansion at quadratic order yielding the *harmonic approximation*

$$(\text{Potential Energy}) \approx U_0 + \frac{1}{2} \sum_{i,j=1}^{N} \sum_{\alpha, \gamma = x, y, z} (s_{i\alpha} - s_{i\alpha}^{(0)})$$
$$\times (s_{j\gamma} - s_{j\gamma}^{(0)}) k_{i\alpha j\gamma},$$

where $s_{i\alpha}$ is the value of the αth Cartesian coordinate for particle i, $s_{i\alpha}^{(0)}$ is the corresponding equilibrium value, $k_{i\alpha j\gamma}$ is a (force) constant, and U_0 is the zero point energy (the minimum value of the potential energy). Note the absence of a linear term in this formula; it does not appear because the first derivatives of the potential energy are zero at the minimum.

Let us consider the consequences of the fact that in the harmonic approximation, the Hamiltonian is a quadratic function of all the coordinates. It is true that different coordinates are coupled or tangled together through the $DN \times DN$ matrix of force constants. (Here, D denotes the dimensionality, and there are DN coordinates in total.) But since the elements of the force constant matrix, $k_{i\alpha, j\gamma}$, are symmetric, a theorem of linear algebra is applicable that says it is possible to untangle the quadratic function by finding a set of *normal coordinates* or *normal modes*. Each normal mode of a harmonic system is a coordinate that oscillates at a given frequency and independent of all other normal modes. There are DN such coordinates for any harmonic (i.e., quadratic) Hamiltonian. Each normal mode is a linear combination of the original set of coordinates, $\{s_{i\alpha}\}$; and when we adopt normal modes as our coordinates, the total Hamiltonian can be written as a sum of DN independent one-dimensional quadratic (i.e., harmonic oscillator) Hamiltonians. To explicitly identify the normal modes of a particular harmonic system, one must diagonalize a $DN \times DN$ matrix. For some discussion of the procedure, see, for example, McQuarrie's *Statistical Mechanics* (Sec. 11.4) or Hill's *Introduction to Statistical Thermodynamics* (Sec. 5.2). But for the treatment we give here, we need only accept the fact that a quadratic form can be diagonalized.

Therefore, by adopting the harmonic approximation, we know

that the Hamiltonian can be expressed as

$$\mathcal{H} = \sum_{\alpha=1}^{DN} \mathcal{H}_\alpha,$$

where

$$\mathcal{H}_\alpha = \text{harmonic oscillator Hamiltonian}$$
$$\text{with a fundamental frequency } \omega_\alpha.$$

Once again, recall that the eigen energy of a harmonic oscillator with frequency ω is

$$(\tfrac{1}{2} + n)\hbar\omega, \qquad n = 0, 1, 2, \ldots.$$

Thus, we introduce the notion of a *phonon*. Let n_α denote the number of phonons in the phonon state with energy $\hbar\omega_\alpha$. Then the energy of a state of the lattice takes on the occupation number form

$$E_v = \sum_{\alpha=1}^{DN} n_\alpha \hbar\omega_\alpha + E_0,$$

where

$$E_0 = U_0 + \sum_{\alpha=1}^{DN} \tfrac{1}{2}\hbar\omega_\alpha.$$

For convenience, let us take U_0 as the zero of energy. Then the canonical partition function for the lattice becomes

$$Q(\beta, N, V) = \sum_{n_1, n_2, \ldots = 0}^{\infty} \exp\left[-\beta \sum_{\alpha} (\tfrac{1}{2} + n_\alpha)\hbar\omega_\alpha\right].$$

Since the exponential factors into a product of independent terms, this expression gives

$$Q = \prod_{\alpha=1}^{DN} \left(\sum_n \exp\left[-\beta(\tfrac{1}{2} + n)\hbar\omega_\alpha\right]\right).$$

The sum over n is performed using the formula for a geometric series yielding

$$\ln Q = -\sum_{\alpha=1}^{DN} \ln[\exp(\beta\hbar\omega_\alpha/2) - \exp(-\beta\hbar\omega_\alpha/2)].$$

Exercise 4.3 Verify this formula.

The sum over phonon states can be partitioned by introducing

$$g(\omega)\, d\omega = \text{the number of phonon states with frequency}$$
$$\text{between } \omega \text{ and } \omega + d\omega.$$

Then

$$\beta A = \int_0^\infty d\omega \, g(\omega) \ln[\exp(\beta\hbar\omega/2) - \exp(-\beta\hbar\omega/2)].$$

This formula is the starting point for the analysis of thermodynamics for harmonic lattices.

Exercise 4.4 Assume that only one phonon level is appreciably populated

$$g(\omega) = \delta(\omega - \omega_0),$$

and determine the low temperature behavior of a harmonic solid. (This picture of a lattice is called the Einstein model.)

Exercise 4.5* Assume the low frequency modes of a lattice are simple plane waves so that

$$g(\omega) = (ND^2/\omega_0^D)\omega^{D-1}, \qquad \omega < \omega_0,$$
$$= 0, \qquad \omega > \omega_0,$$

is a good approximation, and determine the low temperature behavior of a harmonic solid. (This picture of a lattice is called the Debye model, and the cut-off frequency, ω_0, is called the Debye frequency.)

4.4 Ideal Gases of Real Particles

Bosons

Consider a system of N particles that obeys Bose–Einstein statistics and does not interact. One way of evaluating the thermodynamics of such a system is to evaluate the canonical ensemble partition function, $Q = \sum_v \exp(-\beta E_v)$. Unlike the photon or phonon gases that are composed of massless quasi-particles, the systems we consider from now on are composed of particles that cannot be created or destroyed. Thus, if we want to use the occupation representation of states when performing the sum necessary to calculate the canonical partition function, we must constrain the sum to those states in which the total number of particles is fixed at N:

$$Q = \sum_{n_1, n_2, \ldots, n_j, \ldots} \exp\left[-\beta \sum_j n_j \varepsilon_j\right].$$

such that
$$\sum_j n_j = N$$

Here (as always), ε_j denotes the energy of the jth single particle state. The restriction on the summation in the equation produces a combinatorial problem which can be solved; but the precise solution is not simple. By doing statistical mechanics in the grand canonical ensemble, however, the restricted sum does not appear.

In the grand canonical ensemble the partition function is

$$e^{\beta p V} = \Xi = \sum_\nu e^{-\beta(E_\nu - \mu N_\nu)},$$

where ν denotes a state with N_ν particles and energy E_ν. In terms of occupation numbers

$$\Xi = \sum_{n_1, n_2, \ldots, n_j, \ldots} \exp\left[-\beta \sum_j (\varepsilon_j - \mu)n_j\right].$$

Here, the exponential factors and we obtain

$$\Xi = e^{\beta p V} = \prod_j \left\{\sum_{n_j=0}^{\infty} e^{-\beta(\varepsilon_j - \mu)n_j}\right\} = \prod_j \left\{\frac{1}{1 - e^{\beta(\mu - \varepsilon_j)}}\right\}$$

or

$$\beta p V = \ln \Xi = -\sum_j \ln\left[1 - e^{\beta(\mu - \varepsilon_j)}\right].$$

The average occupation number is

$$\langle n_j \rangle = \frac{\sum_\nu n_j e^{-\beta[E_\nu - \mu N_\nu]}}{\Xi} = \frac{\partial \Xi}{\partial(-\beta \varepsilon_j)}\bigg/ \Xi$$

$$= \frac{\partial \ln \Xi}{\partial(-\beta \varepsilon_j)}.$$

Using this formula with Ξ for the ideal Bose gas we obtain

$$\langle n_j \rangle = \frac{1}{e^{\beta(\varepsilon_j - \mu)} - 1}.$$

Notice the singularity when $\mu = \varepsilon_j$. At this point, $\langle n_j \rangle$ diverges; that is, a macroscopic number of particles pile into the same single particle state. The phenomenon is called Bose condensation, and the condensation is thought to be the mechanism for superfluidity.

Recall that photons and phonons are bosons in the sense that any number can exist in the same single particle state. Thus according to the formula we have just derived, the chemical potential of a phonon in an ideal phonon gas is zero. Similarly, the chemical potential of a photon in an ideal photon gas is zero.

Fermions

We now consider an ideal gas of real Fermi particles. Once again it is much easier to work in the grand canonical ensemble, and this ensemble's partition function is

$$\Xi = \sum_{n_1, n_2, \dots, n_j, \dots = 0}^{1} \exp\left[-\beta \sum_j n_j(\varepsilon_j - \mu)\right].$$

Here we have noted that for fermions, $n_j = 0$ or 1 only. As is always the case for non-interacting particles, the exponential in the summand factors and we obtain

$$\Xi = \prod_j \left[\sum_{n_j=0}^{1} e^{-\beta(\varepsilon_j - \mu)n_j}\right]$$

$$= \prod_j [1 + e^{-\beta(\varepsilon_j - \mu)}],$$

or

$$\beta p V = \ln \Xi = \sum_j \ln [1 + e^{\beta(\mu - \varepsilon_j)}].$$

Once again, the average occupation number is given by $\langle n_j \rangle = \partial \ln \Xi / \partial(-\beta \varepsilon_j)$:

$$\langle n_j \rangle = \frac{e^{\beta(\mu - \varepsilon_j)}}{1 + e^{\beta(\mu - \varepsilon_j)}} = \frac{1}{e^{\beta(\varepsilon_j - \mu)} + 1},$$

which is called the Fermi Distribution.

In summary:

$$\langle n_j \rangle_{\substack{\text{F.D.} \\ \text{B.E.}}} = \frac{1}{e^{\beta(\varepsilon_j - \mu)} \pm 1}.$$

Information about correlations between different particles can be described with the averages $\langle n_i n_j \cdots \rangle$. To be specific, let us consider a system of identical fermions. Here, n_i is either zero or one, and $\langle n_i \rangle$ is the probability that a particle is in single particle state i. Similarly

$$\langle n_i n_j \rangle = \text{joint probability that a particle is in}$$
$$\text{state } i \text{ and a particle is in state } j$$

and

$$g_{ij} = \langle n_i n_j \rangle - \langle n_i \rangle \delta_{ij} = \text{joint probability that a particle is in}$$
state i and *another* particle is in state j.

Exercise 4.6 Deduce this last assertion. [*Hint*: Express n_i as a sum of "occupation variables" for each particle in the many particle fermion system.]

Exercise 4.7 For an ideal gas of identical fermions, determine g_{ij} as a functional of $\langle n_i \rangle$.

4.5 Electrons in Metals

As an illustration, we now consider the thermal properties of conducting electrons in metals. To a good approximation, we can model these electrons as an ideal gas of fermions because at high enough densities, the potential energy of interaction between identical fermions is often of little importance. The reason is that since no two identical and therefore indistinguishable fermions can exist in the same state, a high density system will necessarily fill many single particle energy levels. The lowest energy of unoccupied states will have a kinetic energy many times that of $k_B T$, and it is excitations into these states that produce the fluctuations associated with observed finite temperature thermodynamic properties. Thus, when the density of a many particle system of fermions is high enough, the energetics of interactions between particles becomes negligible.

As we see shortly, the conduction electrons of most metals satisfy the criterion of high density. If we assume the conduction electrons are an ideal gas, the average number of electrons occupying the jth single particle state is

$$\langle n_j \rangle = F(\varepsilon_j),$$

where $F(\varepsilon)$ is the Fermi function

$$F(\varepsilon) = [e^{\beta(\varepsilon - \mu)} + 1]^{-1},$$

and ε_j is the energy of the jth single particle state:

$$\varepsilon_j = (\hbar^2 k^2 / 2m),$$

with m denoting electron mass, and the wavevector, $\mathbf{k}$, is quantized according to

$$\mathbf{k} = (\hat{\mathbf{x}} n_x + \hat{\mathbf{y}} n_y + \hat{\mathbf{z}} n_z) \pi / L, \qquad n_\alpha = 0, 1, 2, \ldots,$$

where $L^3 = V$ is the (cubic) volume of the material containing the electrons. These are the standard electron in a box formulas. Notice that the state index, j, must specify the quantum numbers, n_x, n_y, and n_z. In addition, j must specify the spin state (up or down) of the electron. Since $\langle N \rangle = \sum_j \langle n_j \rangle$, we have

$$\langle N \rangle = 2 \int_0^\infty d\varepsilon \, \rho(\varepsilon) F(\varepsilon),$$

where the factor of 2 accounts for the degeneracy of the two spin states, and $d\varepsilon \, \rho(\varepsilon)$ is the number of *structureless* single particle states with energy between ε and $\varepsilon + d\varepsilon$. Equivalently,

$$\langle N \rangle = 2 \int_0^\infty dn_x \int_0^\infty dn_y \int_0^\infty dn_z \, F[\varepsilon(k)]$$

$$= 2 \iiint_0^\infty [dk_x \, dk_y \, dk_z / (\pi/L)^3] F[\varepsilon(k)]$$

$$= \frac{2V}{(2\pi)^3} \iiint_{-\infty}^\infty dk_x \, dk_y \, dk_z \, F[\varepsilon(k)]$$

$$= [2V/(2\pi)^3] \int d\mathbf{k} F[\varepsilon(k)],$$

where we have noted that for large enough volumes V, the spectrum of wavevectors is a continuum so that an integration is appropriate, and the last equality simply introduces a compact notation for integrating over all **k**-space.

Exercise 4.8* Use the Euler–Maclaurin series to prove that the errors incurred by regarding the discrete sums over n_α as integrals are negligible for large V.

To proceed, consider the form of the Fermi function. At $T = 0$,

$$F(\varepsilon) = 1, \qquad \varepsilon < \mu_0,$$
$$= 0, \qquad \varepsilon > \mu_0,$$

where μ_0 is the chemical potential of the ideal electron gas at $T = 0$; it is often called the Fermi energy. The Fermi momentum, p_F, is defined by

$$\mu_0 = p_F^2/2m = \hbar^2 k_F^2/2m.$$

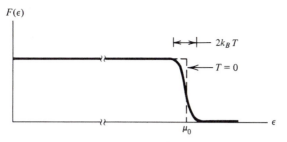

Fig. 4.2. The Fermi function.

Thus, at $T = 0$, we can perform the integral over $F(\varepsilon)$ to obtain $\langle N \rangle$:

$$\langle N \rangle = [2V/(2\pi)^3]\tfrac{4}{3}\pi k_F^3.$$

A typical metal, Cu, has a mass density of 9 g/cm³. By assuming each atom donates an electron to the conducting electron gas, we obtain from this density

$$\mu_0/k_B \approx 80,000°\text{K},$$

which verifies that even at room temperature, the ideal gas approximation is accurate.

Exercise 4.9 Show that the number and conclusion cited here are correct.

Figure 4.2 provides a sketch of the Fermi function for temperatures much lower than μ_0/k_B (e.g., room temperature). Its derivative is a delta-like function, as illustrated in Fig. 4.3.

We can exploit this behavior when computing thermodynamic properties. For example,

$$\langle E \rangle = \sum_j \langle n_j \rangle \varepsilon_j$$

$$= 2 \int_0^\infty d\varepsilon \, \rho(\varepsilon) F(\varepsilon) \varepsilon$$

$$= -\int_0^\infty d\varepsilon \, \Phi(\varepsilon)(dF/d\varepsilon),$$

where in the last equality, we integrated by parts (the boundary term is zero) and introduced

$$\Phi(\varepsilon) = \int_0^\varepsilon dx \, 2\rho(x)x.$$

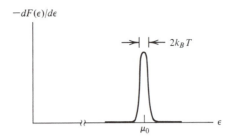

Fig. 4.3. Derivative of the Fermi function.

Since $dF/d\varepsilon$ is highly localized near $\varepsilon = \mu_0$ and $\Phi(\varepsilon)$ is regular, we can profitably expand $\Phi(\varepsilon)$ about $\varepsilon = \mu_0$ and obtain

$$\langle E \rangle = -\sum_{m=0}^{\infty} \frac{1}{m!} \left[\frac{d^m \Phi}{d\varepsilon^m} \right]_{\varepsilon=\mu_0} \int_0^{\infty} \left[\frac{dF}{d\varepsilon} \right] (\varepsilon - \mu_0)^m \, d\varepsilon$$

$$= (\text{constant}) + (k_B T)^2 (\text{another constant}) + O(T^4).$$

Exercise 4.10 Verify this result. [*Hint*: Introduce the transformation of variables $x = \beta(\varepsilon - \mu)$ and note the μ and μ_0 are close at small T.]

Thus, we predict that for a conducting material at low enough temperatures the heat capacity is linear in the temperature; that is,

$$C_v \propto T$$

This prediction has been verified by many experiments.

We have already commented on our neglect of electron-electron interactions. We have also neglected interactions between the lattice atoms and the electrons. That is, we have neglected electron-phonon interactions. It turns out that these subtle interactions are responsible for the phenomenon of superconductivity.

4.6 Classical Ideal Gases, the Classical Limit

We now consider what happens to the statistical behavior of ideal quantum mechanical gases as we approach high temperatures. This is the classical limit. The number of particles is given by $N = \sum_j n_j$. The average number of particles is given by

$$\langle N \rangle = \sum_j \langle n_j \rangle = \sum_j [e^{\beta(\varepsilon_j - \mu)} \pm 1]^{-1},$$

where in the last equality, the upper sign is for Fermi–Dirac statistics and lower for Bose–Einstein statistics. The average density $\langle N \rangle / V$ is the thermodynamic density. When the temperature is high (β is small) and the density is low, many more single particle states are accessible than there are particles. Thus the relation $\langle N \rangle = \sum_j \langle n_j \rangle$ implies that for the considered case, each $\langle n_j \rangle$ must be small; that is, $\langle n_j \rangle \ll 1$. This condition combined with the Fermi–Dirac and Bose–Einstein distributions implies

$$e^{\beta(\varepsilon_j - \mu)} \gg 1. \tag{a}$$

Note that if this equation is true for all ε_j, then $-\beta\mu \gg 1$ when $\beta \to 0$ and $\rho \to 0$. The limiting case of $\beta \to 0$ and $\rho \to 0$ corresponds to the limit of a classical ideal gas.

By virtue of the severe inequality, we have

$$\langle n_j \rangle = e^{-\beta(\varepsilon_j - \mu)} \tag{b}$$

in the classical limit. The chemical potential, μ, is determined by the condition

$$\langle N \rangle = \sum_j \langle n_j \rangle = \sum_j e^{-\beta(\varepsilon_j - \mu)} = e^{\beta\mu} \sum_j e^{-\beta\varepsilon_j},$$

or

$$e^{\beta\mu} = \frac{\langle N \rangle}{\sum_j e^{-\beta\varepsilon_j}}. \tag{c}$$

Thus, combining (b) and (c), we find

$$\langle n_j \rangle = \langle N \rangle \frac{e^{-\beta\varepsilon_j}}{\sum_j e^{-\beta\varepsilon_j}}$$

which is the familiar form of the *classical Boltzmann factor*. Indeed, note that since $\langle n_j \rangle$ is the average number of particles in the jth single particle state, we have

$$\frac{\langle n_j \rangle}{N} = \text{probability of finding a particle in single particle state } j$$
$$\propto e^{-\beta\varepsilon_j}.$$

Now, let's evaluate the canonical partition function in the classical limit. The canonical partition function is related to the Helmholtz free energy by

$$-\beta A = \ln Q,$$

while the grand canonical partition is related to $\beta p V$ by

$$\beta p V = \ln \Xi.$$

Recall that $A = \langle E \rangle - TS$, and the Gibbs free energy is $G = \mu \langle N \rangle = \langle E \rangle + pV - TS$. Using these relations, we obtain

$$\ln Q(\langle N \rangle, V, T) = -\beta \mu \langle N \rangle + \ln \Xi.$$

Inserting the grand partition functions for ideal fermions or bosons into this equation yields

$$\ln Q(\langle N \rangle, V, T) = -\beta \mu \langle N \rangle \pm \sum_j \ln \left[1 \pm e^{\beta(\mu - \varepsilon_j)}\right],$$

where the upper sign is for Fermi particles and the lower for Bose. Using the inequality (a) and the expansion of the logarithm $\ln (1 + x) = x + \cdots$, the above equation becomes

$$\ln Q(\langle N \rangle, V, T) = -\beta \mu \langle N \rangle + \sum_j e^{\beta(\mu - \varepsilon_j)}.$$

Inserting (c) into this formula, we have

$$\ln Q(\langle N \rangle, V, T) = -\beta \mu \langle N \rangle + \langle N \rangle.$$

Next, note that the logarithm of (c) gives

$$\beta \mu = \ln \langle N \rangle - \ln \sum_j e^{-\beta \varepsilon_j}$$

so that

$$\ln Q = -N \ln N + N + N \ln \sum_j e^{-\beta \varepsilon_j}$$

where we have replaced $\langle N \rangle$ by N. Next, we use Stirling's approximation $\ln N! = N \ln N - N$ (which is exact in the thermodynamic limit) to obtain

$$Q = \frac{1}{N!} \left[\sum_j e^{-\beta \varepsilon_j} \right]^N$$

in the classical limit. The factor of $(N!)^{-1}$ reflects the fact that the particles are indistinguishable. It is the only remnant of quantum statistics in the classical limit.

   This formula for the partition function of the classical ideal gas can be derived by an alternative route that does not make use of our previous analysis of the fermion and boson gases. In particular, consider an assembly of N indistinguishable and uncorrelated particles. (In quantum theory, even when particles do not interact, indistinguishability implies correlations due to the required symmetrization of wavefunctions.) If we ignore for the moment the indistinguishability, the partition function is a single particle partition function, q, raised to the Nth power, q^N. As discussed in the introductory remarks of this chapter, however, this result overcounts states. The reason is that there are $N!$ different ways to assign the same set of distinct single particle state labels to the N particles. But each way is equivalent since the particles are indistinguishable. Hence, the result should be $(N!)^{-1}q^N$, which corrects for the overcounting.

4.7 Thermodynamics of an Ideal Gas of Structureless Classical Particles

We will now use the result of the classical limit and consider a system of structureless particles of mass m in a volume V. The energy is due solely to translational center-of-mass motion. The single particle energy can be expressed as

$$\varepsilon_{\mathbf{k}} = \frac{\hbar^2 k^2}{2m}, \qquad \mathbf{k} = \frac{\pi}{L}(n_x\hat{\mathbf{x}} + n_y\hat{\mathbf{y}} + n_z\hat{\mathbf{z}}).$$

Here $L = V^{1/3}$; n_x, n_y, n_z go from 1 to ∞. ($\varepsilon_{\mathbf{k}}$ is the particle in a three-dimensional box energy level.) The classical partition function is then

$$Q(N, V, T) = (N!)^{-1}\left[\sum_{n_x,n_y,n_z=1}^{\infty} \exp\left(-\beta\hbar^2 k^2/2m\right)\right]^N.$$

In the classical limit, $\beta\hbar$ is small, and in the thermodynamic limit L is large. Thus, the difference between consecutive terms in the summand is small, and the sum can be taken over into an integral:

$$\Delta n_x = \frac{L}{\pi}\Delta k_x \rightarrow \frac{L}{\pi}dk_x,$$

$$\Delta n_y = \frac{L}{\pi}\Delta k_y \rightarrow \frac{L}{\pi}dk_y,$$

$$\Delta n_z = \frac{L}{\pi}\Delta k_z \rightarrow \frac{L}{\pi}dk_z,$$

and

$$\sum_{n_x,n_y,n_z} \rightarrow \frac{L^3}{(\pi)^3}\int_0^{\infty} dk_x \int_0^{\infty} dk_y \int_0^{\infty} dk_z.$$

Thus,

$$\sum_{n_x, n_y, n_z} \exp(-\beta \hbar^2 k^2/2m) \rightarrow \frac{V}{\pi^3} \int_0^\infty dk_x \, dk_y \, dk_z$$

$$\times \exp[-\beta \hbar^2 (k_x^2 + k_y^2 + k_z^2)/2m]$$

$$= \frac{V}{(2\pi)^3} \int_{-\infty}^\infty dk_x \, dk_y \, dk_z$$

$$\times \exp[-\beta \hbar^2 (k_x^2 + k_y^2 + k_z^2)/2m].$$

Defining the variable $\mathbf{p} = \hbar \mathbf{k}$, we have

$$Q = (N!)^{-1} \left[\frac{V}{(2\pi)^3 \hbar^3} \int d\mathbf{p} \, e^{-\beta p^2/2m} \right]^N$$

where $d\mathbf{p}$ denotes $dp_x \, dp_y \, dp_z$. The integral can be done in many ways. The result is

$$e^{-\beta A} = Q(N, V, T) = \frac{V^N}{N! h^{3N}} \left(\frac{2\pi m}{\beta} \right)^{3N/2}.$$

The internal energy, $\langle E \rangle$, and the pressure, p, are determined in the usual way:

$$\langle E \rangle = \left(\frac{\partial \ln Q}{\partial(-\beta)} \right)_V = \frac{3}{2} \frac{N}{\beta} = \tfrac{3}{2} N k_B T,$$

$$\beta p = \left(\frac{\partial \ln Q}{\partial V} \right)_\beta = \frac{N}{V}, \quad \text{or} \quad pV = N k_B T.$$

Experimentally, the equation of state for a classical dilute gas is found to be $pV = nRT$ where R is the gas constant and n the number of moles. Our result says that $pV = N k_B T$. This is why, as already stated in Sec. 3.6, the arbitrary constant k_B has the numerical value $k_B = R/N_0 = 1.35805 \times 10^{-16}$ erg/deg, where N_0 is Avogadro's number.

The reader may find it useful to contrast this derivation of the classical ideal gas law with the development carried out in Sec. 3.6. In the earlier derivation, we did not explicitly invoke a classical limit; we assumed only that different particles were uncorrelated. Does this mean that non-interacting quantal fermions or bosons obey the equation of state $pV = nRT$? The answer is no. Indeed, see Exercises 4.18–4.20. The reason is that, even in the absence of interparticle interactions, the quantum mechanical nature of particle indistinguishability implies the existence of interparticle correlations. Thus, while the correlations between different occupation numbers may vanish with Bose–Einstein and Fermi–Dirac statistics, the inter*particle* correlations among indistinguishable particles remain nontrivial. These correlations vanish, however, in the classical limit.

4.8 A Dilute Gas of Atoms

As a further illustration, we now consider a gas of atoms and attempt to account for the internal structure of these particles. We specify the state of an atom with

$$j = (\mathbf{k}, n, v).$$

state of nucleus

electronic state

center-of-mass translation

The center-of-mass translations are uncoupled from the internal structure (described with n and v), and to a good approximation, the nuclear and electronic degrees of freedom are uncoupled (so that n and v are approximate good quantum numbers). Thus

$$\sum_j e^{-\beta \varepsilon_j} = \underbrace{\sum_{\mathbf{k}} \exp\left(-\beta \hbar^2 k^2 / 2m\right)}_{q_{\text{trans}}(T, V)} \underbrace{\sum_{n,v} \exp\left(-\beta \varepsilon_{nv}\right)}_{q_{\text{int}}(T)},$$

where ε_{nv} represents the energy for internal state (n, v). [Note: $q_{\text{int}}(T)$ is said to be independent of the volume V. Why is this?] Let ε_{00} denote the ground energy. Then

$$q_{\text{int}}(T) = e^{-\beta \varepsilon_{00}} \sum_{n,v} \exp\left[-\beta(\varepsilon_{nv} - \varepsilon_{00})\right].$$

For many atoms, internal excitation energies are very large compared to $k_B T$. (For example, 1 eV corresponds to $k_B T$ when $T \approx 10,000°K$.) For this situation, the only terms contributing significantly to the sum are those with the same energy as ε_{00}. Hence

$$q_{\text{int}}(T) \approx e^{-\beta \varepsilon_{00}} \times (\text{degeneracy of ground level})$$
$$= e^{-\beta \varepsilon_{00}} g_0^{(\text{nuc})} g_0^{(\text{elec})}$$

where $g_0^{(\text{nuc})}$ and $g_0^{(\text{elec})}$ are the ground level degeneracies of the nuclear states and electronic states, respectively. If we assume that only spin degrees of freedom are important in the nucleus, then

$$g_0^{(\text{nuc})} = (2I + 1),$$

where I is the total spin quantum number of the nucleus. Combining all of these formulas leads to

$$-\beta A = \ln\left[(N!)^{-1}q_{\text{trans}}^N(T, V)q_{\text{int}}^N(T)\right]$$
$$= -\beta N \varepsilon_{00} + N \ln\left[g_0(2I + 1)\right] + \ln\left[(N!)^{-1}q_{\text{trans}}^N(T, V)\right].$$

The factor involving $q_{\text{trans}}(T, V)$ was analyzed in the previous example when we studied structureless particles. Notice that the internal structure affects the energy and entropy of the gas, but it leaves the pressure unaltered.

Before leaving our discussion of atoms, the following puzzle is worth thinking about: The electronic energy levels of a hydrogen atom go as

$$\varepsilon_n = -(\varepsilon_0/n^2), \qquad n = 1, 2, \ldots, \infty.$$

Thus,

$$q_{\text{int}}(T) = g_1 e^{\beta \varepsilon_0} + g_2 e^{\beta \varepsilon_0/4} + g_3 e^{\beta \varepsilon_0/9} + \cdots + g_n e^{\beta \varepsilon_0/n^2} + \cdots,$$

where g_n is the degeneracy of the $(n-1)$st electronic energy level. Clearly, $g_n \geq 1$, and for large enough n, the nth term is simply g_n. Thus, the series is divergent! How can this difficulty be resolved?

Exercise 4.11 Answer the above question. [*Hint*: Consider the average spatial extent of the electronic wave function for the hydrogen atom as a function of n.]

4.9 Dilute Gas of Diatomic Molecules

We now consider the thermal properties of a gas composed of molecular species. The internal energetics of molecules involve vibrational and rotational motions as well as electronic and nuclear spin degrees of freedom. To treat these motions, we use the *Born–Oppenheimer* approximation. In this approximation, one imagines solving Schrödinger's equation for the molecule with the nuclei nailed down. For each electronic state, there will be a different energy for each nuclear configuration. The electronic energy as a function of nuclear coordinates forms an effective (electronically averaged) potential for the nuclei (this is the Hellman–Feynman theorem). The potential is called a *Born–Oppenheimer surface*. For each electronic state, there is a different surface. The rotational and vibrational states of the nuclei are determined by solving Schrödinger's equation for the nuclei in these effective potentials.

For this scheme to be accurate, it must be true that the electronic motion is fast compared to the nuclear motion so that nuclear kinetic energy may be neglected when determining the electronic wavefunction. It can be shown that the Born–Oppenheimer approximation is exact in the limit $m_e/M \to 0$, where m_e is the mass of an electron, and M is the mass of a nucleus. Of course, the ratio is never zero, but it is small enough that the approximation is often accurate. It must also be true that no two Born–Oppenheimer energies are close to one another in any accessible nuclear configuration. If the Born–Oppenheimer surfaces do get close, the small perturbations due to the nuclear motions (i.e., the nuclear kinetic energy) will cause a breakdown in the approximation.

In the Born–Oppenheimer approximation, the wavefunction of a diatomic molecule is

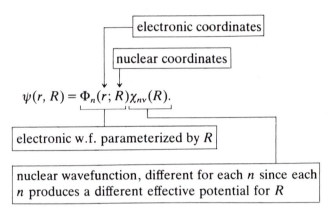

With this form, the prescription described above is: (1) Calculate $\Phi_n(r, R)$ from

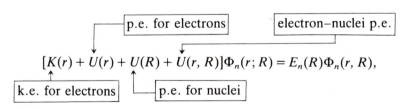

(2) Determine $\chi_{nv}(R)$ from

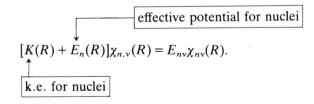

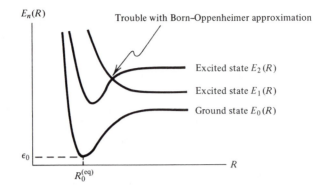

Fig. 4.4. Hypothetical Born-Oppenheimer potentials for a diatomic molecule.

Exercise 4.12 Verify that this procedure provides an accurate solution to Schrödinger's equation for the molecule provided the kinetic energy (k.e.) of the electrons is much larger than the k.e. of the nuclei. Why is there "trouble" at the indicated spot in Fig. 4.4.

Figure 4.4 is a schematic picture of some electron averaged potentials (i.e., Born–Oppenheimer surfaces).

With the Born–Oppenheimer approximation we write

$$q_{\text{int}}(T) = \sum_{n,v} \langle n, v| \exp\left[-\beta \mathcal{H}_{\text{int}}\right] |n, v\rangle$$

$$= \sum_{v} \{\langle \chi_{0v}| \exp\left[-\beta \mathcal{H}_{\text{eff}}^{(0)}(R)\right] |\chi_{0v}\rangle$$

$$+ \langle \chi_{1v}| \exp\left[-\beta \mathcal{H}_{\text{eff}}^{(1)}(R)\right] |\chi_{1v}\rangle + \cdots\},$$

where $\mathcal{H}_{\text{eff}}^{(n)}(R)$ is the effective (electronically averaged) Hamiltonian for the nuclear coordinates, R, when the molecule is the nth electronic state; that is, $\mathcal{H}_{\text{eff}}^{(n)}(R) = K(R) + E_n(R)$. If we neglect all but the ground electronic state,

$$q_{\text{int}}(T) \approx g_0 e^{-\beta \varepsilon_0} \sum_{v} \exp\left[-\beta (E_{0v} - \varepsilon_0)\right].$$

Further, let us assume that vibrational and rotational motions of the nuclei are uncoupled. This is often a good approximation due to a mismatch in time scales (vibrations are usually more rapid than rotations). Hence, we shall model the nuclear motions in the effective potential $E_0(R)$ as harmonic oscillating rigid rotor motions. Then

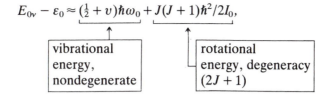

$v = 0, 1, 2, \ldots,$ and $J = 0, 1, 2, \ldots.$ Here, ω_0 and I_0 are the fundamental frequency and moment of inertia, respectively, associated with the ground state Born–Oppenheimer potential. That is, $I_0 = \mu_0(R^{(eq)})^2$ and

$$\omega_0^2 = \mu^{-1}[\partial^2 E_0(R)/\partial R^2]_{R=R_0(eq)},$$

where $\mu = [(1/m_A) + (1/m_B)]^{-1}$ is the reduced mass of the two nuclei.

At this point, it looks as if we should just sum over J and v to get $q_{int}(T)$. However, there is a subtlety when the molecule is homonuclear. In that case, one must be careful to sum over only those states that are odd or even on interchanging nuclei depending on whether the nuclei are fermions or bosons, respectively. In the classical limit, this restriction leads to a division by a symmetry number which accounts for the overcounting of states when one does distinguish between the configurations 1—2 and 2—1. In our treatment below, we will use this classical result. For a discussion of the quantum mechanics, see for example, McQuarrie's *Statistical Mechanics* (Sec. 6.5) or Hill's *Introduction to Statistical Thermodynamics* (Sec. 22.8).

Thus, we obtain

$$q_{int}(T) \approx g_0 e^{-\beta \varepsilon_0} \underbrace{(2I_A + 1)(2I_B + 1)}_{} q_{rot} q_{vib} / \sigma_{AB},$$

where

$$q_{rot}(T) = \sum_{J=0}^{\infty} (2J + 1) \exp\left[-J(J + 1)\beta \hbar^2 / 2I_0\right]$$

and

$$q_{vib}(T) = \sum_{v=0}^{\infty} \exp\left[-(\tfrac{1}{2} + v)\beta\hbar\omega_0\right].$$

The vibrational contribution, $q_{vib}(T)$, is computed with the aid of the geometric series,

$$q_{vib}(T) = [\exp(\beta\hbar\omega_0/2) - \exp(-\beta\hbar\omega_0/2)]^{-1}.$$

Exercise 4.13 Derive this result.

The rotational contribution is more difficult. We can convert the sum to an integral with the Euler–Maclaurin series provided the spacing between rotational levels is small compared to $k_B T$. This condition is usually met (except for hydrogen). Thus

$$q_{rot}(T) \approx \int_0^{\infty} dJ(2J+1) \exp\left[-J(J+1)\beta\hbar^2/2I_0\right] = T/\theta_{rot},$$

where the last equality follows from a change in integration variables from J to $J(J+1)$, and θ_{rot} is the rotational temperature, $\hbar^2/2I_0 k_B$.

Exercise 4.14* Use the Euler–Maclaurin series to show that

$$q_{rot}(T) = \frac{T}{\theta_{rot}}\left[1 + \frac{1}{3}\left(\frac{\theta_{rot}}{T}\right) + \frac{1}{15}\left(\frac{\theta_{rot}}{T}\right)^2 + \cdots\right].$$

We can combine these results with

$$q_{trans}(T, V) = V\left[\frac{2\pi M}{\beta h^2}\right]^{3/2}, \qquad M = m_A + m_B,$$

and thus

$$
\begin{aligned}
-[\beta A(N, V, T)]_{\text{diatomic ideal gas}} \\
\approx -N\ln N + N + N\ln V \\
+ \tfrac{3}{2}N\ln(2\pi M k_B T/h^2) \\
+ N\ln q_{int}(T).
\end{aligned}
$$

with

$$q_{int}(T) \approx g_0 e^{-\beta\varepsilon_0}(2I_A + 1)(2I_B + 1)\sigma_{AB}^{-1}$$
$$\times (T/\theta_{rot})[\exp(\beta\hbar\omega_0/2) - \exp(-\beta\hbar\omega_0/2)]^{-1}.$$

These equations provide a theory for the thermodynamic properties of molecular gases.

4.10 Chemical Equilibria in Gases

Among the many possible applications, the formulas we have derived
for gas phase partition functions can be used to compute chemical
equilibrium constants. But first it is convenient to establish a notation
and thermodynamic preliminaries. Consider the reaction

$$aA + bB \rightleftarrows cC + dD.$$

with annotations: "molecular species" pointing to the species, and "stoichiometric coefficient" pointing to the coefficients.

We can write this expression more compactly as

$$0 = \sum_{i=1}^{4} v_i X_i,$$

where v_i is the reaction coefficient. That is, $v_1 = c$, $v_2 = d$, $v_3 = -a$,
$v_4 = -b$, $X_1 = C$, $X_2 = D$, $X_3 = A$, $X_4 = B$. The stoichiometry places
a restriction on changes in mole numbers:

$$\Delta n_A = (b/a)^{-1} \Delta n_B = -(c/a)^{-1} \Delta n_C = -(d/a)^{-1} \Delta n_D.$$

Thus, the first-order virtual displacement of the internal energy due
to variations in mole numbers is

$$\delta E = \sum_{i=1}^{4} \mu_i \, \delta n_i$$
$$= \delta n_A [\mu_A + (b/a)\mu_B - (c/a)\mu_C - (d/a)\mu_D].$$

As a result, the equilibrium condition $(\Delta E)_{S,V,n} > 0$ implies

$$0 = \mu_A + (b/a)\mu_B - (c/a)\mu_C - (d/a)\mu_D,$$

or, more generally

$$0 = \sum_{i=1}^{r} v_i \mu_i$$

when components $1, 2, \ldots, r$ are involved in an equilibrium chemical
reaction.

This condition tells us a great deal about the composition of
reactants and products at equilibrium. To see why, let us define the
quantity γ_i by the equation

$$\beta \mu_i = \ln \rho_i \gamma_i, \qquad \rho_i = N_i / V.$$

Then at chemical equilibrium

$$0 = \sum_{i=1}^{r} \nu_i \ln \rho_i \gamma_i$$

$$= \ln \prod_{i=1}^{r} (\rho_i \gamma_i)^{\nu_i}.$$

Thus

$$K \equiv \prod_{i=1}^{r} (\rho_i)^{\nu_i} = \prod_{i=1}^{r} (\gamma_i^{-1})^{\nu_i},$$

which is called the *law of mass action*.

To make further progress we must determine a molecular expression for μ_i and thus γ_i. In the classical ideal gas approximation, the partition function of an r-component mixture is

$$Q(\beta, V, N_1, \ldots, N_r) = \frac{1}{N_1!} \frac{1}{N_2!} \cdots \frac{1}{N_r!} q_1^{N_1} q_2^{N_2} \cdots q_r^{N_r},$$

where $q_i = q_i(\beta, V)$ is the single particle partition function. Thus

$$\beta A = -\ln Q = \sum_{i=1}^{r} [\ln N_i! - N_i \ln q_i].$$

As a result,

$$\beta \mu_i = (\partial \beta A / \partial N_i)_{\beta, V, N_j}$$

$$= \ln N_i - \ln q_i.$$

However,

$$q_i = (V/\lambda_i^3) q_i^{(\text{int})},$$

where $q_i^{(\text{int})}$ is for species i the $q_{\text{int}}(T)$ of the previous section, and

$$\lambda_i = h/\sqrt{2\pi m_i k_B T}$$

is the "thermal wavelength" for a particle of mass m_i. Thus

$$\beta \mu_i = \ln [\rho_i (\lambda_i^3 / q_i^{(\text{int})})].$$

Hence, we have identified γ_i, and the equilibrium constant is

$$K = \prod_{i=1}^{r} [q_i^{(\text{int})}/\lambda_i^3]^{\nu_i}.$$

The equilibrium constant is a function of temperature since both $q_i^{(\text{int})}$ and λ_i are functions of temperature.

This last formula is one of the early triumphs of statistical mechanics as it represents a molecular expression for the equilibrium constant appearing in the law of mass action.

Additional Exercises

4.15. Consider an isomerization process

$$A \rightleftarrows B,$$

where A and B refer to the different isomer states of a molecule. Imagine that the process takes place in a dilute gas, and that $\Delta \varepsilon$ is the energy difference between state A and state B. According to the Boltzmann distribution law, the equilibrium ratio of A and B populations is given by

$$\langle N_A \rangle / \langle N_B \rangle = (g_A/g_B)e^{-\beta \Delta \varepsilon},$$

where g_A and g_B are the degeneracies of states A and B, respectively. Show how this same result follows from the condition of chemical equilibria, $\mu_A = \mu_B$.

4.16. Consider the system described in Exercise 4.15. The canonical partition function is

$$Q = \frac{1}{N!} q^N,$$

where N is the total number of molecules, and q is the Boltzmann weighted sum over all single molecule states, both those associated with isomers of type A and those associated with isomers of type B.

(a) Show that one may partition the sum and write

$$Q = \sum_P \exp\left[-\beta A(N_A, N_B)\right]$$

with

$$-\beta A(N_A, N_B) = \ln \left[(N_A! N_B!)^{-1} q_A^{N_A} q_B^{N_B}\right],$$

where $\sum_P$ is over all the partitions of N molecules into N_A molecules of type A and N_B molecules of type B, q_A is the Boltzmann weighted sum over states of isomer A, and q_B is similarly defined.

(b) Show that the condition of chemical equilibria is identical to finding the partitioning that minimizes the Helmholtz free energy

$$(\partial A / \partial \langle N_A \rangle) = (\partial A / \partial \langle N_B \rangle) = 0,$$

subject to the constraint that $\langle N_A \rangle + \langle N_B \rangle = N$ is fixed.

4.17. For the system described in Exercises 4.15 and 4.16, we have the canonical partition function

$$Q = \sum_{\substack{N_A, N_B \\ (N_A + N_B = N)}} q_A^{N_A} q_B^{N_B} / N_A! N_B!$$
$$= (q_A + q_B)^N / N!.$$

Show that

$$\langle N_A \rangle = q_A (\partial \ln Q / \partial q_A)_{q_B, N}$$
$$= N q_A / (q_A + q_B).$$

Use this result and the analogous formula for $\langle N_B \rangle$ to show that $\langle N_A \rangle / \langle N_B \rangle = q_A / q_B$. Next, consider the fluctuations from these mean values. Express the average of the square fluctuation, $(N_A - \langle N_A \rangle)^2$, as appropriately weighted sums over states, and show that

$$\langle [N_A - \langle N_A \rangle]^2 \rangle = q_A (\partial \langle N_A \rangle / \partial q_A)_{q_B, N}$$
$$= \langle N_A \rangle \langle N_B \rangle / N.$$

Derive a similar expression for the fluctuations in N_B.

4.18. (a) Show that for an ideal gas of structureless fermions, the pressure is given by

$$\beta p = \frac{1}{\lambda^3} f_{5/2}(z),$$

where $z = \exp(\beta \mu)$,

$$\lambda = (2\pi \beta \hbar^2 / m)^{1/2},$$

m is the mass of the particle,

$$f_{5/2}(z) = \frac{4}{\sqrt{\pi}} \int_0^\infty dx \, x^2 \ln(1 + z e^{-x^2})$$
$$= \sum_{l=1}^\infty (-1)^{l+1} z^l / l^{5/2},$$

and the chemical potential is related to the average density,

$$\rho = \langle N \rangle / V,$$

by

$$\rho \lambda^3 = f_{3/2}(z) = \sum_{l=1}^\infty (-1)^{l+1} z^l / l^{3/2}.$$

(b) Similarly, show that the internal energy, $\langle E \rangle$, obeys the relation

$$\langle E \rangle = \tfrac{3}{2} p V.$$

4.19. Consider the ideal Fermi gas of Exercise 4.18 in the *high temperature* and/or *low density* regime ($\rho \lambda^3 \ll 1$).

(a) Show that

$$z = \rho \lambda^3 + (\rho \lambda^3)^2 / 2\sqrt{2} + \cdots.$$

(b) Use this result together with the Fermi distribution for $\langle n_p \rangle$ to deduce the Maxwell–Boltzmann distribution

$$\langle n_p \rangle \approx \rho \lambda^3 e^{-\beta \varepsilon_p},$$

where p stands for momentum and $\varepsilon_p = p^2/2m$.

(c) Show that the thermal wavelength, λ, can be viewed as an average De Broglie wavelength since

$$\lambda \sim h / \langle |p| \rangle.$$

(d) Show that

$$\beta p / \rho = 1 + \rho \lambda^3 / (2)^{5/2} + \cdots.$$

Why does a finite value of $\rho \lambda^3$ lead to deviations from the classical ideal gas law? Why should you expect the quantum deviations to vanish when $\rho \lambda^3 \to 0$?

4.20. Consider now the ideal Fermi gas of Exercise 4.18 at *low temperature* and/or *high density* ($\rho \lambda^3 \gg 1$).

(a) Show that

$$\rho \lambda^3 = f_{3/2}(z) \approx (\ln z)^{3/2} 4/3\sqrt{\pi},$$

hence

$$z \approx e^{\beta \varepsilon_F},$$

where

$$\varepsilon_F = (\hbar^2/2m)(6\pi^2 \rho)^{2/3}.$$

[*Hint:* Use the integral representation of

$$f_{3/2}(z) = (4/\sqrt{\pi}) \int_0^\infty dx \, x^2 (z^{-1} e^{x^2} + 1)^{-1}.]$$

(b) Show that

$$p = 2\varepsilon_F \rho / 5 [1 + O(k_B^2 T^2 / \varepsilon_F^2)].$$

Hence the pressure does not vanish at $T = 0$. Why?

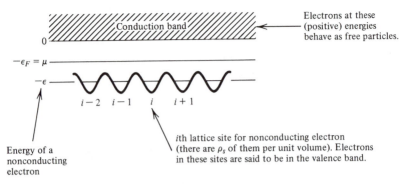

Electrons at these (positive) energies behave as free particles.

$i-2$ $i-1$ i $i+1$

ith lattice site for nonconducting electron (there are ρ_s of them per unit volume). Electrons in these sites are said to be in the valence band.

Energy of a nonconducting electron

Fig. 4.5. Model for semiconductors.

4.21. In this problem, consider a model for semiconductors based upon the Fermi ideal gas approximation for electrons. Our schematic view of the system is shown in Fig. 4.5. Both ε_F and $\varepsilon - \varepsilon_F$ are large compared to $k_B T$.

(a) Show that the average electron density, $\rho = \langle N \rangle / V$, is given by

$$\rho = 2(\rho_s)[e^{-\beta(\varepsilon - \varepsilon_F)} + 1]^{-1}$$
$$+ [2/(2\pi)^3] \int d\mathbf{k}\,[e^{\beta(\varepsilon_k + \varepsilon_F)} + 1]^{-1},$$

where $\varepsilon_k = \hbar^2 k^2 / 2m_e$. The first of these two terms represents the average concentration of electrons in the lattice sites, and the second term represents the concentration in the conduction band.

(b) Noting that $\beta \varepsilon_F \gg 1$ and $\beta(\varepsilon - \varepsilon_F) \gg 1$, show that

$$pn = (\rho_s / \lambda^3) 4 e^{-\beta \varepsilon},$$

where λ is the thermal wavelength for an electron, p is the average density of unfilled lattice sites, and n is the density of electrons in the conduction band. This relationship provides a law of mass action for semiconductors since the right-hand side plays the role of an equilibrium constant for the product of variable concentrations on the left.

4.22. Consider an ideal gas of argon atoms in equilibrium with argon atoms adsorbed on a plane surface. Per unit surface area, there are ρ_s sites at which the atoms can be adsorbed, and when they are, their energy is $-\varepsilon$ per adsorbed atom. Assume the particles behave classically (is this a good approximation at

1 atm and 300°K?) and show that

$$(\rho_{ad}/\rho_g) = \rho_s \lambda^3 e^{\beta \varepsilon},$$

where λ is the thermal wavelength for an argon atom, $\rho_g = \beta p$ is the gas phase density, and ρ_{ad} is the number of atoms adsorbed per unit area. [*Hint*: At phase equilibrium, chemical potentials are equal. Further, as in Sec. 4.10, $\beta \mu$ in the gas phase is $\ln \rho \lambda^3$.] Note the similarity between this classical result and that found in Exercise 4.21. Comment.

4.23. (a) Prove that if the energy eigenvalues of a system can be expressed as a sum of independent contributions $E = E_A + E_B + E_C$ (e.g., electronic energy, vibrational energy, rotational energy) that the heat capacity can be written $C_V = C_V^{(A)} + C_V^{(B)} + C_V^{(C)}$. In addition, show that the heat capacity is independent of zero point energy.

(b) Derive an expression for the electronic heat capacity assuming that there are only three significant electronic states and that they have energies and degeneracies given by ε_0, g_0; ε_1, g_1; ε_2, g_2.

(c) Given that the energies required for electronic transitions correspond roughly to u.v. light ($\sim$50,000°K), show how the calculated room temperature heat capacity of a diatomic molecule will change if the electronic degrees of freedom are totally neglected. What if the ground electronic state degeneracy is included but all the excited electronic states are neglected?

(d) Show how the room temperature *entropy* of the same molecule will change in these two cases.

4.24. State all the necessary assumptions and calculate the entropy and C_p (in units of cal per mole-deg) for HBr at 1 atm pressure and 25°C given that $\hbar \omega / k_B = 3700$°K and $\hbar^2 / 2Ik_B = 12.1$°K. Here $\hbar \omega$ is the energy spacing between the ground and the first excited vibrational state in HBr and I is the moment of inertia of HBr in the ground vibrational and electronic states. The ground electronic state for HBr is nondegenerate.

4.25. Use the information compiled in Chapter 8 of Hill's *Introduction to Statistical Thermodynamics* to calculate the equilibrium constant, K, for the reaction

$$I_2 \rightleftarrows 2I$$

when the reaction occurs in the dilute gas phase at $T = 1000$°K.

4.26. Consider a one-component gas of non-interacting classical structureless particles of mass m at a temperature T.

    (a) Calculate exactly the grand canonical partition function, Ξ, for this system as a function of volume, V, temperature, and chemical potential, μ. Your result should look like

$$\Xi = \exp(zV),$$

    where z is a function of T and μ.

    (b) From the result of part (a), determine the pressure, p, as a function of T and the average particle density, ρ.

    (c) For 1 cc of gas at STP, compute the relative root mean square of the density fluctuations, $[\langle(\delta\rho)^2\rangle/\rho^2]^{1/2}$.

    (d) Calculate the probability of observing a spontaneous fluctuation in 1 cc of gas at STP for which the instantaneous density differs from the mean by one part in 10^6.

Bibliography

The material covered in this chapter is discussed in all elementary texts. Those referenced at the end of Chapter 3 may be useful to the student.

The discussion of the classical limit—the transition from quantum to classical statistical mechanics—deserves a more complete treatment than provided herein or in other introductions to the subject. Perhaps the best treatments employ Feynman's path integral formulation of quantum theory. A most useful introduction to path integrals in statistical mechanics is given in Chapter 3 of Feynman's text:

    R. P. Feynman, *Statistical Mechanics* (Benjamin, Reading, Mass., 1972).

The discussion of symmetrized many particle wavefunctions and occupation numbers touches on the subject of "second quantization," which is a useful notational scheme. This subject is treated in Chapter 6 of Feynman's text.

CHAPTER 5

Statistical Mechanical Theory of Phase Transitions

In Chapter 4 we considered many examples of ideal gases—systems in which particles or degrees of freedom do not interact with each other. We now leave that simple class of models and consider the situation in which interparticle interactions cause correlations between many particles. Phase transitions provide the most striking manifestations of interparticle interactions. Our discussion of the microscopic theory for this type of phenomena will focus on a simple class of lattice models. Most of what we say, however, has broad implications extending far beyond these particular systems. In particular, in the context of the lattice models we will discuss the meaning of order parameters and broken symmetry, and we will introduce the generally applicable approaches to treating coupled or highly correlated systems with analytical theory: mean field theory and renormalization group theory. These powerful concepts and techniques play a central role in all of the physical sciences.

5.1 Ising Model

We consider a system of N spins arranged on a lattice as illustrated in Fig. 5.1. In the presence of a magnetic field, H, the energy of the system in a particular state, v, is

$$E_v = -\sum_{i=1}^{N} H\mu s_i + \text{(energy due to interactions between spins)},$$

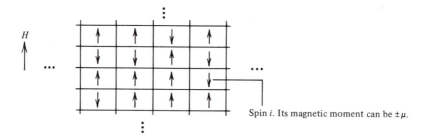

Spin i. Its magnetic moment can be $\pm\mu$.

Fig. 5.1. Spins on a lattice.

where $s_i = \pm 1$. A simple model for the interaction energy is

$$-J \sum_{ij}' s_i s_j,$$

where J is called a coupling constant, and the primed sum extends over nearest-neighbor pairs of spins. The spin system with this interaction energy is called the *Ising model*.

Notice that when $J > 0$, it is energetically favorable for neighboring spins to be aligned. Hence, we might anticipate that for low enough temperature, this stabilization will lead to a *cooperative phenomenon* called *spontaneous magnetization*. That is, through interactions between nearest-neighbors, a given magnetic moment can influence the alignment of spins that are separated from the given spin by a macroscopic distance. These *long ranged correlations* between spins are associated with a *long ranged order* in which the lattice has a net magnetization even in the absence of a magnetic field. The magnetization

$$\langle M \rangle = \sum_{i=1}^{N} \mu s_i$$

in the absence of the external magnetic field, H, is called the spontaneous magnetization.

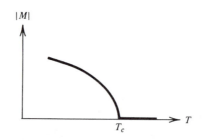

Fig. 5.2. Spontaneous magnetization.

Unless the temperature is low enough (or J is high enough), there will be no net magnetization. Let T_c (the Curie temperature or critical temperature) denote the highest temperature for which there can be non-zero magnetization. We expect a curve like the one illustrated in Fig. 5.2.

Exercise 5.1 Show that for positive J and zero field (i.e., $H \rightarrow 0^+$), the lowest energy of the Ising model with a cubic lattice is

$$E_0 = -DNJ, \qquad D = \text{dimensionality},$$

and that this lowest energy corresponds to all the spins aligned (either all up or all down). [*Hint*: The number of nearest-neighbors for any given spin is 2, 4, or 6 in 1, 2, or 3 dimensions, respectively.] What is the ground state energy for a two-dimensional Ising model with a triangular lattice?

Exercise 5.2 Determine the magnitude of the spontaneous magnetization of an Ising model at zero temperature.

Provided $T_c > 0$, the system undergoes an order-disorder transition: a phase transition. We study this magnetic system because it is simpler than fluids. However, we will see that the magnetic transition is closely analogous to liquid-gas transitions, and many other processes of interest.

The macroscopic properties of this magnetic lattice can be computed from the partition function

$$Q(\beta, N, H) = \sum_v e^{-\beta E_v}$$

$$= \sum_{s_1} \sum_{s_2} \cdots \sum_{s_N = \pm 1} \exp \left[\beta \mu H \sum_{i=1}^{N} s_i + \beta J \sum_{ij}' s_i s_j \right].$$

The interaction term couples the different s_i's so that the multiple sums are tangled together. In a one-dimensional lattice

$$
\begin{array}{ccccccc}
\uparrow & \uparrow & \downarrow & \cdots & \uparrow & \downarrow & \downarrow & \cdots & \uparrow \\
1 & 2 & 3 & & i-1 & i & i+1 & & N
\end{array}
$$

the interaction energy can be reduced to a sum over a single index

$$-J \sum_{i=1}^{N} s_i s_{i+1}$$

(where we have used periodic boundary conditions; that is, the $N + 1$ spin is the first spin). For that case, the partition function can be evaluated analytically yielding (at zero field)

$$Q(\beta, N, 0) = [2 \cosh(\beta J)]^N.$$

Exercise 5.3 Verify that this result is correct for large N.

It is also quickly verified that the one-dimensional model predicts no spontaneous magnetization at any finite temperature (see Exercise 5.21).

The physical reason for the absence of spontaneous magnetization is easily understood by considering the energetics of excitations to disordered states. For example, one of the two ground states

$$\uparrow \quad \uparrow \quad \uparrow \cdots \uparrow \quad \uparrow \cdots \uparrow$$

$$1 \quad 2 \qquad \frac{N}{2} \qquad N$$

has an energy $-NJ$ and a net magnetization per particle of μ. This is an ordered state. The disordered state

$$\uparrow \uparrow \uparrow \cdots \uparrow \quad \downarrow \cdots \downarrow$$

$$1 \quad 2 \qquad \frac{N}{2} \qquad N$$

with magnetization of *zero* has energy $(-N + 4)J$. This small change in energy, only one part in N, is insufficient to stabilize the ordered state. Indeed, for a one-dimensional system, the net magnetization per particle should vanish for all temperatures higher than $T \sim J/Nk_B$, which is vanishingly small for macroscopic N.

In a two-dimensional system, however, the excitation energy to a disordered state is much higher. For example, the energy for the configuration

$$\cdots \uparrow \uparrow \uparrow \uparrow \uparrow \cdots$$
$$\uparrow \uparrow \uparrow \uparrow \uparrow$$
$$\cdots \downarrow \downarrow \downarrow \downarrow \downarrow \cdots$$
$$\downarrow \downarrow \downarrow \downarrow \downarrow$$

is $N^{1/2}$ parts out of N higher than the energy of the perfectly ordered configuration. This difference in energy can be sufficient to stabilize an ordered state.

In fact, in two and three dimensions, the Ising model does exhibit an order-disorder phase transition. However, the demonstration of this fact is nontrivial and represents one of the major achievements of 20th century science. In the 1940s, Lars Onsager showed that the partition function of the two-dimensional Ising model (at zero field) is

$$Q(\beta, N, 0) = [2 \cosh(\beta J)e^I]^N,$$

where

$$I = (2\pi)^{-1} \int_0^\pi d\phi \, \ln\{\tfrac{1}{2}[1 + (1 - \kappa^2 \sin^2 \phi)^{1/2}]\}$$

with

$$\kappa = 2 \sinh(2\beta J)/\cosh^2(2\beta J).$$

Onsager's result implies that the free energy associated with this partition function is nonanalytic. Further, it can be shown that a spontaneous magnetization exists for all temperatures below

$$T_c = 2.269J/k_B.$$

[T_c is the solution to $\sinh(2J/k_B T_c) = 1$.] Near this temperature, Onsager found that the heat capacity, $C = (\partial \langle E \rangle / \partial T)_{H=0}$, is singular:

$$(C/N) \sim (8k_B/\pi)(\beta J)^2 \ln|1/(T - T_c)|.$$

Further, the behavior of the magnetization is

$$(M/N) \sim (\text{constant})(T_c - T)^\beta, \qquad T < T_c,$$

where $\beta = 1/8$. (Don't confuse this exponent with the reciprocal temperature.)

No one has solved the three-dimensional Ising model analytically. But numerical solutions have established that a critical temperature exists which is roughly twice the value for two dimensions. Near that temperature,

$$(C/N) \propto |T - T_c|^{-\alpha}$$

and

$$(M/N) \propto (T_c - T)^\beta, \qquad T < T_c,$$

where the *critical exponents* have the values

$$\alpha \approx 0.125, \qquad \beta \approx 0.313.$$

We will soon consider two approximate schemes for accounting for the interactions between particles, and see what predictions these methods make with regard to the phase transition in the Ising magnet. First, however, there are a few more things that should be said about the general physics of phase transitions.

5.2 Lattice Gas

Here we will show that with a simple change in variables, the Ising magnet becomes a model for density fluctuations and liquid-gas phase transformations. To begin, we construct a model based upon the lattice drawn at the beginning of this chapter. In this case, however, the lattice divides space into cells. Each cell can be either occupied by a particle or unoccupied. We will let $n_i = 0$ or 1 denote the occupation number for the ith cell. The upper bound of $n_i = 1$ is effectively an excluded volume condition saying that no pair of particles can be closer than the lattice spacing. Attractions between neighboring particles are accounted for in this model by saying that when particles are in nearest-neighbor cells, the energy associated with each such pair is $-\varepsilon$. The total energy for a given set of occupation numbers is then

$$-\varepsilon \sum_{i,j}{}' n_i n_j.$$

We will neglect any further detail concerning the configurations of particles in the system. In this approximation, the configuration of the system is determined by the set $\{n_i\}$, and the grand canonical partition function is given by

$$\Xi = \sum_{\substack{n_1,\dots,n_N \\ =0,1}} \exp\left\{\beta\mu \sum_{i=1}^{N} n_i + \beta\varepsilon \sum_{ij}{}' n_i n_j\right\},$$

where N is the number of cells (not particles) and μ is the chemical potential. The volume of the system is N times the volume of a cell.

The model system with this partition function is called the lattice gas. It is isomorphic with the Ising magnet. The correspondence is established by making the change of variables

$$s_i = 2n_i - 1.$$

One finds that "spin up" in the Ising magnet corresponds to an occupied cell in the lattice gas, "spin down" corresponds to an empty cell, the magnetic field maps (within constants) to the chemical potential, and the coupling constant J in the Ising magnet is $\varepsilon/4$ in

---

Exercise 5.4 Make the correspondence concrete. In particular, derive the precise relationships between the parameters in the Ising magnet and those in the lattice gas that make the canonical partition function of the former identical within a proportionality constant to the grand partition function of the latter.

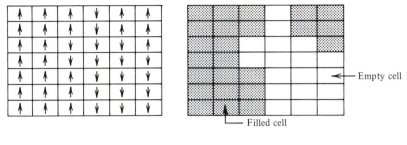

Ising magnet   Lattice gas

Fig. 5.3. Isomorphic systems.

the lattice gas. In Fig. 5.3, we illustrate the correspondence for one particular configuration.

More sophisticated lattice models for density fluctuations can be constructed by generalizing the nature of the lattice and the number of components and by increasing the number of states per lattice site from the two state (up and down spins) Ising magnet. Multistate generalizations are often called *Potts models,* and the use of complicated lattices are often called *decorated lattice models.*

5.3 Broken Symmetry and Range of Correlations

One feature of the order-disorder phenomenon in the Ising magnet should cause all but the most casual observer to pause. In the absence of the magnetic field, the model is symmetric with regard to the up and down directions of the spin. Indeed, the ground state with all spins aligned is two-fold degenerate since the total alignment can be either up or down. Therefore, in the absence of an external magnetic field, it would seem that an exact statistical mechanical calculation of the magnetization through the formula

$$\langle M \rangle = Q^{-1} \sum_{v} \left(\sum_{i=1}^{N} \mu s_i \right) e^{-\beta E_v}$$

would necessarily give zero. The reasoning here is quite simple: for every configuration with positive $M_v = \sum_i \mu s_i$, symmetry demands that there is an equally weighted configuration with negative M_v. Hence, the sum is zero. How then should we think about the broken symmetry of spontaneous magnetization?

One answer is constructed by considering a free energy that depends upon the magnetization. In particular, imagine summing over all states v for which the net magnetization is constrained to the

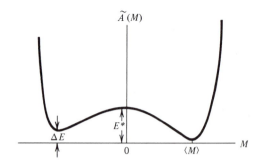

Fig. 5.4. Reversible work function for the magnetization.

value M; that is,

$$\tilde{Q}(M) = \sum_{v} \Delta(M - M_v)e^{-\beta E_v},$$

where $\Delta(M - M_v)$ is 1 when $M = M_v$, and zero otherwise. Clearly, $Q = \sum_M \tilde{Q}(M)$ and $\tilde{Q}(M)/Q$ is the probability for observing the system with magnetization M. The quantity

$$-k_B T \ln \tilde{Q}(M) = \tilde{A}(M)$$

is the free energy that determines the reversible work required to change the magnetization of the system. From our discussion of the energetics associated with destroying long ranged order, we imagine that when the system is below the critical temperature, $\tilde{A}(M)$ appears as drawn schematically in Fig. 5.4. The energy ΔE is due to the coupling of the spins to the external magnetic field. (We imagine that the field is very small, perhaps so small that it is only slightly bigger than a field proportional to N^{-1}. Otherwise, the scale of this highly schematic figure is surely not correct.) As $H \to 0^+$, the bias for up spins disappears and $\tilde{A}(M)$ becomes symmetric with positive and negative values of M equally weighted. The energy E^* is an activation energy. If the system is in a state with M close to $\langle M \rangle$, one requires a fluctuation with energy of the size of E^* in order for the system to reach a state with magnetization near $-\langle M \rangle$.

As we have discussed, E^* is very large in two and three dimensions, scaling as $N^{1/2}$ and $N^{2/3}$, respectively, when $H = 0$. It is a surface tension energy (or line tension in two dimensions). Thus, due to the Boltzmann weight of configurations, the likelihood of fluctuations between $\langle M \rangle$ and $-\langle M \rangle$ states becomes vanishingly small in the limit of a large system. As a result, the occurrence of spontaneous magnetization or the symmetry breaking resulting from the long ranged correlations can be viewed as follows: Through the application of an external field, the system is prepared with a magnetization of a particular sign. The field can then be made arbitrarily weak, and if we are below the critical temperature, $E^* \neq 0$, and spontaneous fluctuations will not be of sufficient size to destroy the broken symmetry.

Exercise 5.5 Consider this type of discussion in the context of liquid-gas phase equilibria and gravitational fields. Use the lattice gas model as a concrete example.

The fluctuating magnetization, *M*, is the *order parameter* for the Ising system. Here, the words "order parameter" are used to denote a fluctuating variable the average value of which provides a signature of the order or broken symmetry in the system. For the corresponding lattice gas, the order parameter is the deviation of the density from its critical point value. To expand on the meaning of order parameters, it is useful to introduce another concept: the *range of correlations*—that is, the distance over which fluctuations in one region of space are correlated or affected by those in another region. If two points are separated by a distance larger than that range, then the different fluctuations at these two points will be uncorrelated from each other.

Consider the situation in which the range of correlations, *R*, is a microscopic distance—that is, no larger than a few lattice spacings. In that case, the lattice can be partitioned into a huge number of statistically independent cells. Each cell would have a side length *L* that is significantly larger than *R* but still microscopic in size. See Fig. 5.5. The net magnetization in each cell is uncorrelated to that of its neighbors. Hence, there is no macroscopic cooperativity, and the average total magnetization will be zero.

If, on the other hand, *R* was macroscopic in size, a net average magnetization in a macroscopic sample could then exist. We see, therefore, that the broken symmetry associated with a finite value of average order parameter is intimately related to the existence of *long range order*—that is, a range of correlations macroscopic in size.

To write equations associated with these words, we introduce the pair correlation function between spins *i* and *j*,

$$c_{ij} = \langle s_i s_j \rangle - \langle s_i \rangle \langle s_j \rangle.$$

For the corresponding lattice gas model, c_{ij} is a constant times the correlation function between densities in cells *i* and *j*. According to its definition, c_{ij} vanishes when the spin (or occupation number) at

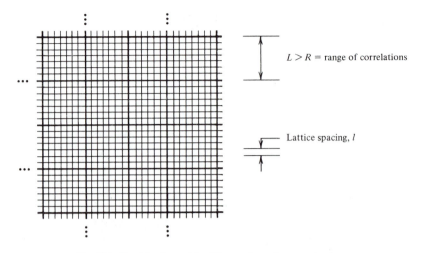

$L > R$ = range of correlations

Lattice spacing, l

Fig. 5.5. Partitioning with different lengths on a lattice.

lattice site i is uncorrelated with that at lattice site j. Hence,

$$\sum_{j=2}^{N} c_{1j} \approx \text{number of spins correlated to spin 1.}$$

We have singled out spin 1 for notational convenience. All the spins, of course, are equivalent. Note that as the range of correlations grows, the number computed in the above summation also grows.

This number of correlated spins and thus the spatial range of correlations is related to the susceptibility,

$$\chi = \frac{1}{N} \left(\frac{\partial \langle M \rangle}{\partial \beta H} \right)_{\beta}.$$

To establish the relationship we note that from the analysis of fluctuations in magnetization, we have the general result

$$\chi(\beta, H) = \frac{1}{N} \langle (\delta M)^2 \rangle,$$

where

$$\delta M = M - \langle M \rangle$$
$$= \mu \sum_{i=1}^{N} [s_i - \langle s_i \rangle].$$

Therefore,

$$\chi = (\mu^2/N) \sum_{i,j=1}^{N} [\langle s_i s_j \rangle - \langle s_i \rangle \langle s_j \rangle]$$

$$= \mu^2 \sum_{j=1}^{N} c_{1j},$$

where, in the second equality, we have used the definiton of c_{ij} and the fact that all lattice sites are equivalent. According to this last formula, the divergence of the susceptibility is associated with the existence of *long ranged correlations*. The reasoning is that the right-hand side counts numbers of correlated spins, and this number increases as the range of correlations increases.

One way χ can diverge is related to the phenomenon of symmetry breaking. Here, two phases coexist at $T < T_c$, and the application of an infinitesimal field will suffice to produce a net macroscopic alignment of spins. In particular, for positive H, if $N \rightarrow \infty$ and then $H \rightarrow 0^+$ for $T < T_c$, then $\langle M \rangle = Nm_0\mu$, where $m_0\mu$ is the spontaneous magnetization per spin (the subscript zero emphasizes that the field is sent to zero). On the other hand, if $H < 0$ and $N \rightarrow \infty$ and then $H \rightarrow 0^-$, then $\langle M \rangle = -Nm_0\mu$. Hence, for $T < T_c$, the $\langle M \rangle$ of an infinite system is a discontinuous function of the field at $H = 0$. Therefore, its derivative, $\partial \langle M \rangle / \partial H$ is divergent at $H = 0$.

To examine this behavior in more detail, consider the case where $T < T_c$, N is big (but not yet infinite), and $H = 0$. Due to the symmetry of the system

$$\langle s_i \rangle = 0.$$

At the same time*

$$\sum_{j=1}^{N} \langle s_1 s_j \rangle = Nm_0.$$

Now given this result, we see that the susceptibility at $H = 0$ for large N and $T < T_c$ is

$$\chi = Nm_0\mu^2.$$

The behavior implied by the formula is depicted in Fig. 5.6. As the diagram and the last formula indicate, the divergence of χ that we associate with symmetry breaking and long range order is a divergence that appears in the limit of a very large system. Recall also that

---

* One may understand this equality by noting that the value of the tagged spin, s_1, is sufficient to bias the system into one of the two states of broken symmetry (spin up or spin down). That is, if $s_1 = +1$, then $\sum_j \langle s_1 s_j \rangle$ is Nm_0; and if $s_1 = -1$, the sum is then $-Nm_0 s_1 = +Nm_0$. Finally, the probability of $s_i = \pm 1$ is 1/2 in both cases.

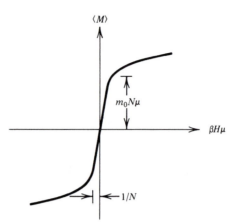

Fig. 5.6. The average magnetization for $T < T_c$ and large but finite N.

χ is proportional to the mean square fluctuations in the magnetization. The divergence of χ is therefore connected to the presence of macroscopic fluctuations. In other words, the possibility of broken symmetry implies correlations in fluctuations over macroscopic distances. The phenomenon we are considering here is like the behavior of a coalesced drop of liquid that in the absence of gravity would wander through the vapor. The large macroscopic fluctuations are simply the appearance and disappearance of the droplet in the region of observation. See Fig. 5.7. Notice that these fluctuations are quenched by the application of a small symmetry breaking field.

At or near the critical point, however, a somewhat different situation occurs. Here too χ diverges. But now it is because of the disappearance of the distinction between two phases (e.g., spin up and spin down, or high density and low density). In other words, when considering the bistable potential shown in Fig. 5.4, the barrier vanishes; that is, the surface energy tends to zero. In that case, $(\partial H / \partial \langle M \rangle)_\beta$ tends to zero as $\beta \to \beta_c$ at $H = 0$. [For a lattice gas model, or any fluid in general, the corresponding relation is

$$(\partial \beta \mu / \partial \rho)_T = \frac{1}{\rho}(\partial \beta p / \partial \rho)_\beta \to 0$$

along the critical isochore as $T \to T_c$ from above.] Since the barrier or distinction between phases ceases to exist at the critical point, the application of a small external field is not sufficient to break symmetry and quench the macroscopic fluctuations implied by $\chi \to \infty$. The critical point is therefore an infinitely susceptible state possessed

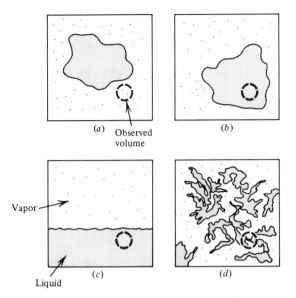

Fig. 5.7. Fluctuations with phase equilibria.

by fluctuations. These fluctuations are not random but highly correlated over large macroscopic distances.

Figure 5.7 illustrates what we might observe for the fluctuations with liquid-vapor equilibrium. Pictures (a) and (b) are two situations with the temperature T well below its critical value, T_c, and no gravitational field. Picture (c) also has $T < T_c$, but a gravitational field is applied. Picture (d) imagines the fluid near its critical point. The observed volume is indicated by the dashed line. The nearest-neighbor spacing, l, in the lattice gas model corresponds to the thickness of the lines drawn in these figures.

5.4 Mean Field Theory

In general, the theoretical treatment of systems undergoing phase transitions requires the use of approximations. The first such approach we consider is a self-consistent field method that has provided the foundation for nearly all many-body theories developed prior to 1970. To illustrate the method, we will apply it to the Ising magnet. The idea is to focus on one particular particle (in this case a spin) in the system and assume that the role of the neighboring particles (spins) is to form an average *molecular* (magnetic) *field* which acts on the tagged particle (spin). See Fig. 5.8. This approach, therefore,

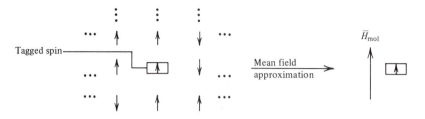

Fig. 5.8. Schematic view of mean field theory.

neglects the effects of fluctuations that extend beyond the length scale associated with the primary or tagged lattice cell. The method includes only those fluctuations that occur within the tagged cell, and since these involve only one particle, the method succeeds at reducing the many body statistical mechanics problem into a few (i.e., one) body problem. Procedures of this type can never be exact, but often they can be very accurate and quite useful. They are least accurate, of course, for systems near critical points where cooperativity in fluctuations extends over large distances.

To begin the mean field analysis, let us write the Ising model energy, E_v,

$$E_v = -\mu H \sum_i s_i - \tfrac{1}{2} \sum_{i,j} J_{ij} s_i s_j,$$

where

$$J_{ij} = J, \quad i \text{ and } j \text{ nearest neighbors},$$
$$= 0, \quad \text{otherwise}.$$

The force exerted on s_i due to the neighboring spins and the external field, H, is

$$-(\partial E_v / \partial s_i) = \mu H + \sum_j J_{ij} s_j.$$

Therefore, the instantaneous field impinging on spin i is given by

$$\mu H_i = \mu H + \sum_j J_{ij} s_j.$$

As the neighboring spins fluctuate, H_i fluctuates from its mean

$$\langle H_i \rangle = H + \sum_j J_{ij} \langle s_j \rangle / \mu$$
$$= H + Jz \langle s_i \rangle / \mu,$$

where in the second equality we have noted that $\langle s_i \rangle = \langle s_j \rangle$ for all i and j in the Ising model, and z is the number of nearest neighbors around a given tagged spin. It is the fluctuations of H_i away from its mean that couples the fluctuations of the tagged spin to those in its environment. Indeed, in the mean field approximation which neglects the deviations of H_i from $\langle H_i \rangle$, the force on a tagged spin is independent of the instantaneous configuration of the neighboring spins.

The statistical mechanics of a system of uncoupled spins in a fixed magnetic field was examined in Chapter 3 (see Exercises 3.18 and 3.19). Using the results derived there, we can compute the average value of a tagged spin, labeled 1 for convenience, in the mean field approximation

$$\langle s_1 \rangle \approx \sum_{s_1=\pm 1} s_1 \exp\left[\beta\mu(H+\Delta H)s_1\right] \bigg/ \sum_{s=\pm 1} \exp\left[\beta\mu(H+\Delta H)s\right],$$

where ΔH is the molecular or environmental contribution to the mean field; that is,

$$\Delta H = Jz\langle s_1 \rangle/\mu.$$

Performing the sums over the two spin states yields

$$m = \tanh(\beta\mu H + \beta zJm), \tag{a}$$

where m is the magnetization per particle per μ:

$$m = \langle M \rangle/N\mu = \left\langle \sum_{i=1}^{N} \mu s_i \right\rangle \bigg/ N\mu$$

$$= \langle s_i \rangle = \langle s_1 \rangle.$$

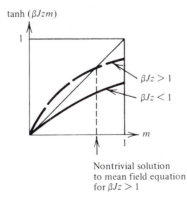

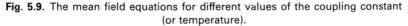

Fig. 5.9. The mean field equations for different values of the coupling constant (or temperature).

Equation (a) for m is a transcendental equation whose solution yields the magnetization predicted by this *self-consistent* mean field theory. It is "self-consistent" in the sense that the mean field which influences the average value of the magnetization depends itself upon the average of the magnetization. Note from Fig. 5.9 that non-zero solutions for m exist for $H = 0$ when $\beta z J > 1$. Thus, for a square lattice we predict the critical temperature

$$T_c = 2DJ/k_B.$$

For $T < T_c$, the solution of $m = \tanh(\beta J z m)$ is

$$\beta = \frac{1}{2Jzm} \ln\left(\frac{1+m}{1-m}\right).$$

Exercise 5.6 Verify this result. By Taylor expansion, analyze the precise form of m for temperatures near $2DJ/k_B$ and show that the critical exponent β in this mean field theory has the value $1/2$. Show that at $T = 0$, the mean field theory for the spontaneous magnetization yields $m = \pm 1$.

Exercise 5.7 Draw a graph of T vs. m as given by the mean field theory. This is the mean field theory prediction of the spontaneous magnetization sketched at the beginning of this chapter.

Exercise 5.8 Show that the total internal energy in the mean field theory is

$$\langle E \rangle = -N\mu Hm - (1/2)JNzm^2.$$

At $H = 0$ and $T = 0$, this gives $\langle E \rangle = -NDJ$, in agreement with the exact result. What does the theory predict for $T > T_c$? Is the prediction correct?

Notice that for one dimension, the mean field theory predicts that an order-disorder transition occurs at a finite temperature $T_c = 2J/k_B$. But the exact analysis of the one-dimensional Ising model yields $T_c = 0$ (i.e., no spontaneous magnetization at finite temperatures). Thus, the theory is "infinitely wrong" in one dimension. Spontaneous fluctuations destroy the order. But mean field theory, which neglects fluctuations, predicts long range order. In two dimensions, the Onsager solution yields $T_c \approx 2.3J/k_B$, while the theory yields $T_c = 4J/k_B$. In three dimensions, the percentage error is smaller: the correct answer (known from numerical work) is $T_c \approx$

$4J/k_B$, whereas the theory gives $T_c = 6J/k_B$. In each case, the neglect of fluctuations predicts transitions to an ordered state at temperatures higher than the true transition temperature.

The theoretical estimates for the critical temperature in two and three dimensions can be significantly improved by making more sophisticated mean field approximations. These improvements are constructed by tagging more than a single spin. For example, one might consider summing over all but a pair of nearest-neighbor spins, s_1 and s_2. Then the molecular field acting on this pair can be approximated as a mean field. This approach reduces the statistical mechanics to a tractable two-body problem, and it accounts for fluctuations at the single particle and pair level, but neglects those involving larger numbers of particles. While it is possible to make significant improvements in this way, mean field theories neglect fluctuations over length scales larger than those associated with a small number of particles, and this neglect will always predict incorrect critical exponents. For example, mean field theories will always yield $\beta = 1/2$, independent of dimensionality. (Actually, in dimensions higher than three, mean field theory can become correct. The reason is not obvious, but can you think of the physical reason for it to be true? The fact that mean field theory increases in accuracy with increasing dimensionality is hinted at by the comparison of critical temperatures.)

5.5 Variational Treatment of Mean Field Theory

In this section we embellish the mean field treatment of the Ising model by introducing the concepts of *thermodynamic perturbation theory* and a variational principle that can be used to optimize first-order perturbation theory. The methods are quite general and serve as a prescription for systematically deriving mean field theories. The embellishment is not necessary, however, for understanding the basis of renormalization group theory as described in the next two sections, and the reader may wish to pass to that section, returning here at a later time.

The Ising model problem is the computation of the Boltzmann weighted sum

$$Q = \sum_{s_1, s_2, \ldots} \exp\left[-\beta E(s_1, s_2, \ldots, s_N)\right],$$

with

$$E(s_1, \ldots, s_N) = -\tfrac{1}{2} \sum_{i,j} J_{ij} s_i s_j - \mu H \sum_i s_i,$$

where J_{ij} is J for nearest neighbors i and j, and zero otherwise. The interaction terms, $-J_{ij}s_is_j$, are responsible for cooperativity, but also for the complexity of computing Q. In the last section, however, we saw that by neglecting fluctuations from the mean in the environment of each spin, the system behaves as one composed of independent entities. Therefore, one way to begin a discussion of mean field theory is to consider an energy function of the form

$$E_{\text{MF}}(s_1, \ldots, s_N) = -\mu(H + \Delta H) \sum_{i=1}^{N} s_i.$$

A system with this energy function is a model with independent spins, and each spin fluctuates under the influence of a static or mean field. This model presents a simple statistical mechanical problem. The partition function is

$$Q_{\text{MF}} = \sum_{\substack{s_1,\ldots,s_N \\ =\pm 1}} \prod_{j=1}^{N} \exp\left[\beta\mu(H + \Delta H)s_j\right]$$

$$= \{2\cosh\left[\beta\mu(H + \Delta H)\right]\}^N, \tag{a}$$

and the average of any of the s_i's is

$$\langle s_1 \rangle_{\text{MF}} = \tanh\left[\beta\mu(H + \Delta H)\right]. \tag{b}$$

In adopting a mean field description, the static molecular field ΔH must be identified. A physical argument was presented in Sec. 5.4 which set $\Delta H = zJ\langle s_1 \rangle/\mu$. The question we now ask is whether this identification was actually the optimum one. In particular, once we adopt the physical picture associated wth $E_{\text{MF}}(s_1, \ldots, s_N)$ (i.e., independent spins fluctuating in an effective static field), we can strive to optimize the parameterization of that model. Of course, in this case, the mean field model is particularly simple being characterized by only one parameter, ΔH.

To optimize, we consider performing a perturbation theory with the mean field model as the reference. This is done as follows: Let

$$\Delta E(s_1, \ldots, s_N) = E(s_1, \ldots, s_N) - E_{\text{MF}}(s_1, \ldots, s_N).$$

Then

$$Q = \sum_{s_1,\ldots,s_N} \exp\left[-\beta(E_{\text{MF}} + \Delta E)\right],$$

where the arguments of $E_{\text{MF}}(s_1, \ldots, s_N)$ and $\Delta E(s_1, \ldots, s_N)$ are understood but omitted for notational simplicity. By factoring the summand, we can also write

$$Q = \sum_{s_1,\ldots,s_N} \exp(-\beta E_{\mathrm{MF}}) \exp(-\beta \Delta E)$$

$$= Q_{\mathrm{MF}} \sum_{s_1,\ldots,s_N} \exp(-\beta E_{\mathrm{MF}}) \exp(-\beta \Delta E)/Q_{\mathrm{MF}}.$$

Now, let us define

$$\langle \cdots \rangle_{\mathrm{MF}} = Q_{\mathrm{MF}}^{-1} \sum_{s_1,\ldots,s_N} [\cdots] \exp(-\beta E_{\mathrm{MF}})$$

as the Boltzmann averaging operation which is weighted with the mean field energy function $E_{\mathrm{MF}}(s_1, \ldots, s_N)$. Then,

$$Q = Q_{\mathrm{MF}} \langle \exp(-\beta \Delta E) \rangle_{\mathrm{MF}}.$$

This last equation, which is an exact factorization of the partition function, is the starting point for thermodynamic perturbation theory. The mean field model is the reference or zeroth-order system, the perturbation energy is $\Delta E(s_1, \ldots, s_N)$, and the effects of this perturbation are computed by performing averages weighted by the Boltzmann factor for the reference system. In developing a mean field theory, we assume that fluctuations about the mean field energy are small; that is, in some sense, ΔE is small. Thus, we Taylor expand the second term:

$$\langle \exp(-\beta \Delta E) \rangle_{\mathrm{MF}} = \langle 1 - \beta \Delta E + \cdots \rangle_{\mathrm{MF}}$$

$$= 1 - \beta \langle \Delta E \rangle_{\mathrm{MF}} + \cdots$$

$$= \exp(-\beta \langle \Delta E \rangle_{\mathrm{MF}}) + \cdots,$$

where the neglected terms are second and higher order in ΔE. We therefore arrive at the first-order perturbation theory result

$$Q \approx Q_{\mathrm{MF}} \exp[-\beta \langle E - E_{\mathrm{MF}} \rangle_{\mathrm{MF}}].$$

How good can this approximation be? The following bound is useful:

$$e^x \geq 1 + x.$$

Exercise 5.9 Draw a graph of e^x vs. x and compare it with $(1+x)$ vs. x. Show that $e^x > 1 + x$ for all real $x \neq 0$. [*Hint*: Note that for real x, e^x is a monotonically increasing function of x as is its derivative too.]

By applying this bound, we have

$$\langle e^f \rangle = e^{\langle f \rangle} \langle e^{(f - \langle f \rangle)} \rangle$$

$$\geq e^{\langle f \rangle} \langle (1 + f - \langle f \rangle) \rangle = e^{\langle f \rangle}.$$

Therefore, in the context of the thermodynamic perturbation theory above, we have

$$Q \geqslant Q_{MF} \exp\left(-\beta\langle E - E_{MF}\rangle_{MF}\right).$$

Exercise 5.10 Derive the first-order perturbation theory for the free energy and show that it is an upper bound to the exact free energy.

This inequality is often called the *Gibbs–Bogoliubov–Feynman bound*. We can use the bound to optimize the mean field reference system since we can adjust the molecular field ΔH so as to maximize the right-hand side. That is, ΔH is determined by solving the equation

$$0 = \frac{\partial}{\partial \Delta H} Q_{MF} \exp\left(-\beta\langle \Delta E\rangle_{MF}\right). \tag{c}$$

The calculation proceeds as follows: First,

$$-\beta\langle \Delta E\rangle_{MF} = \beta N\{\tfrac{1}{2}Jz\langle s_1\rangle_{MF}^2 - \mu\Delta H\langle s_1\rangle_{MF}\}, \tag{d}$$

where we have used the fact that since spins are uncorrelated in the mean field model, $\langle s_i s_j\rangle_{MF} = \langle s_i\rangle_{MF}\langle s_j\rangle_{MF}$ for $i \neq j$, and since all spins are the same on the average, $\langle s_i\rangle_{MF} = \langle s_j\rangle_{MF}$. By combining (d) and (a) with (c), and performing the differentiation, we obtain

$$0 = \beta Jz\langle s_1\rangle_{MF}(\partial\langle s_1\rangle_{MF}/\partial\Delta H) - \beta\mu\Delta H(\partial\langle s_1\rangle_{MF}/\partial\Delta H)$$

or

$$Jz\langle s_1\rangle_{MF} = \mu\Delta H.$$

Exercise 5.11 Verify this result. Further, show that with this choice for ΔH, the first-order perturbation approximation for the free energy corresponds to $Q \approx Q_{MF}$, and that $-k_B T \ln Q_{MF}$ is actually an upper bound to the exact free energy.

This identification of ΔH is identical to that adopted on physical grounds in Sec. 5.4, and by combining it with Eq. (b) for $\langle s_1\rangle_{MF}$, we obtain the same self-consistent mean field equations studied in that section. Those equations are therefore the optimum theory that can be constructed once one adopts the mean field model in which the Ising magnet is pictured as a system of independent spins, each one fluctuating under the influence of an average static environment.

5.6 Renormalization Group (RG) Theory

We now consider a method that can account for large length scale fluctuations, the renormalization group (RG) theory. This approach to understanding phase transitions was developed in 1971 by Kenneth Wilson, who was awarded the 1982 Nobel Prize in Physics for this contribution. Wilson's method is very general and has wide applicability extending well beyond the field of phase transitions. Within this area of research, however, the theory can be viewed as an extension and implementation of phenomenological ideas suggested by Leo Kadanoff in the 1960s.

Several of the concepts in the RG theory can be illustrated with the one-dimensional Ising model. So we start with that system even though it does not exhibit a phase transition. In the absence of the magnetic field, the partition function, Q, is given by

$$Q(K, N) = \sum_{\substack{s_1, s_2, \ldots, s_N \\ = \pm 1}} \exp\left[K(\cdots + s_1 s_2 + s_2 s_3 + s_3 s_4 + s_4 s_5 + \cdots)\right],$$

where

$$K = J/k_B T.$$

The first idea in the RG theory is to remove from the problem a finite fraction of the degrees of freedom by averaging (summing) over them. This is to be contrasted with the mean field theory approach where all but a very few degrees of freedom are removed from explicit consideration. To be specific, let us partition the summand of Q as follows:

$$Q(K, N) = \sum_{s_1, s_2, \ldots, s_N} \exp\left[K(s_1 s_2 + s_2 s_3)\right] \exp\left[K(s_3 s_4 + s_4 s_5)\right] \cdots.$$

Now we can sum over all the even numbered spins, $s_2, s_4, s_6, \ldots$. The result is

$$Q(K, N) = \sum_{s_1, s_3, s_5, \ldots} \{\exp\left[K(s_1 + s_3)\right] + \exp\left[-K(s_1 + s_3)\right]\}$$
$$\times \{\exp\left[K(s_3 + s_5)\right] + \exp\left[-K(s_3 + s_5)\right]\} \cdots.$$

By performing these sums, we have removed every other degree of freedom:

$$\text{O O O O O} \cdots \longrightarrow \text{O} \qquad \text{O} \qquad \text{O} \cdots$$
$$1 \;\; 2 \;\; 3 \;\; 4 \;\; 5 \qquad\qquad\quad 1 \qquad\;\; 3 \qquad\;\; 5$$

The second important idea in the RG theory is to cast this partially summed partition function into a form that makes it look the same as a partition function for an Ising model with $N/2$ spins and (perhaps) a different coupling constant (or reciprocal temperature) K. If this rescaling is possible, then we can develop a recursion relation from which we can compute $Q(K, N)$ starting from a system with another coupling constant (e.g., zero). Thus, we look for a function of K, $f(K)$, and a new coupling constant K' such that

$$e^{K(s+s')} + e^{-K(s+s')} = f(K)e^{K'ss'}$$

for all $s, s' = \pm 1$. If we can find these quantities, then

$$Q(K, N) = \sum_{s_1, s_3, s_5, \ldots} f(K) \exp(K's_1s_3) f(K) \exp(K's_3s_5) \cdots$$
$$= [f(K)]^{N/2} Q(K', N/2),$$

which would be the desired recursion relation. A transformation like this is called a *Kadanoff transformation*.

To determine the quantities K' and $f(K)$, we note that if $s = s' = \pm 1$, then

$$e^{2K} + e^{-2K} = f(K)e^{K'}.$$

The only other possibility is $s = -s' = \pm 1$ from which we have

$$2 = f(K)e^{-K'}$$

or

$$f(K) = 2e^{K'}.$$

Thus, we have two equations for two unknowns. The solution is

$$K' = (1/2) \ln \cosh (2K),$$
$$f(K) = 2 \cosh^{1/2} (2K). \tag{a}$$

Armed with these formulas, consider

$$\ln Q = Ng(K).$$

Within a factor of $-k_BT$, $Ng(K)$ is a free energy, and since free energies are extensive, we expect $g(K)$ to be intensive—that is, independent of system size. From the recursion relation, $\ln Q(K, N) = (N/2)\ln f(K) + \ln Q(K', N/2)$, we have $g(K) = (1/2)\ln f(K) + (1/2)g(K')$, or since $f(K) = 2 \cosh^{1/2} (2K)$,

$$g(K') = 2g(K) - \ln [2\sqrt{\cosh (2K)}]. \tag{b}$$

Equations (a) and (b) are called renormalization group (RG) equations. (They describe transformations which obey the group property, and they provide a renormalization scheme.) If the partition function is known for one value of K', we can generate $\ln Q = Ng(K)$ for other values with this recursion or "renormalization." Notice that in the renormalization obtained from (a) and (b), the new coupling constant, K', as computed from (a) is always less than K.

An alternative set of RG equations would be

$$K = (1/2) \cosh^{-1}(e^{2K'}), \tag{c}$$

which is the inverse of (a), and

$$g(K) = (1/2)g(K') + (1/2)\ln 2 + K'/2, \tag{d}$$

which is obtained by noting $f(K) = 2\exp(K')$.

Exercise 5.12 Derive these RG equations, and show that $K > K'$.

To see how these equations work, we will apply (c) and (d) starting with a small value of the coupling constant. Repeated application will generate $g(K)$ at successively larger values of the coupling constant. Let's start with $K' = 0.01$. For such a small coupling constant, interactions between spins are nearly negligible. Thus, $Q(0.01, N) \approx Q(0, N) = 2^N$. As a result,

$$g(0.01) \approx \ln 2.$$

We now start the iteration. From (c) and (d) we find

$$K = 0.100\,334,$$

$$g(K) = 0.698\,147.$$

We now use these numbers as the new K primed quantities and obtain

$$K = 0.327\,447,$$

$$g(K) = 0.745\,814,$$

and so on.

	K	Renormalization group	Exact
	0.01	ln 2	0.693 197
	0.100 334	0.698 147	0.698 172
	0.327 447	0.745 814	0.745 827
Successive	0.636 247	0.883 204	0.883 210
application	0.972 710	1.106 299	1.106 302
of RG	1.316 710	1.386 078	1.386 080
equations	1.662 637	1.697 968	1.697 968
(c) and (d)	2.009 049	2.026 876	2.026 877
	2.355 582	2.364 536	2.364 537
	2.702 146	2.706 633	2.706 634

Notice how each iteration gives an even more accurate result. What would happen if we moved in the opposite direction?

Exercise 5.13 Start with $K = 10$, which is large enough to approximate $Q(10, N) \approx Q(K \to \infty, N) = 2 \exp(NK)$—that is, $g(10) \approx 10$. Apply the RG equations (a) and (b) and generate a table like that on p. 141 but which progresses from large K to small K. Show that by applying equations (c) and (d), the errors in the nth iteration are 2^{-n} smaller than the error in the initial estimate of g; and show that the errors *grow* exponentially when you apply (a) and (b).

The successive application of the RG equations can be represented by a flow diagram. Each iteration using Eq. (c) leads to higher values of K:

$$\times\!\!\longrightarrow\quad\longrightarrow\quad\longrightarrow\quad\longrightarrow\!\!\times$$

$K = 0$ $\qquad\qquad\qquad\qquad\qquad\qquad\qquad$ $K = \infty$

Each iteration using Eq. (a) leads to lower values of K:

$$\times\!\!\longleftarrow\quad\longleftarrow\quad\longleftarrow\quad\longleftarrow\!\!\times$$

$K = 0$ $\qquad\qquad\qquad\qquad\qquad\qquad\qquad$ $K = \infty$

There are two points, $K = 0$ and $K = \infty$, for which the recursion does not change K. These values of K are called *fixed points*. The fact that there is uninterrupted flow between $K = 0$ and $K = \infty$ (i.e., there are no fixed points at finite K) means that there is no possibility in the one-dimensional Ising model for a phase transition.

On removing degrees of freedom, we transform the problem to a nearly identical one with a larger length scale. In this one-dimensional example, the removal of degrees of freedom leads to a smaller coupling constant K. We can understand this phenomenon physically since there is no long range order (except at $T = 0$, i.e., $K \to \infty$), and hence longer length scales should be associated with less order, and thus a smaller K. By removing degrees of freedom we move to smaller K and thereby transform the problem to a weak coupling one in which K is close to zero. Near a *trivial fixed point*, such as $K = 0$, properties are easily computed from perturbation theory.

Notice that at $K = 0$ and ∞, the fixed points for this system, the lattice is completely disordered or ordered, respectively. When completely ordered, the system appears to be the same, independent of the length scale with which it is viewed. A similar statement is true

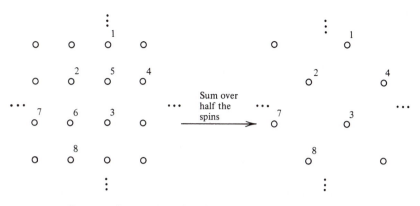

Fig. 5.10. Decimation of half the spins on a square lattice.

for the entirely disordered case. The invariance to a length scale transformation is an essential feature of the RG fixed points, even when they are the trivial $K = 0$ and ∞ fixed points.

For systems exhibiting a phase transition such as the two-dimensional Ising magnet, we will find nontrivial fixed points associated with phase transitions.

5.7 RG Theory for the Two-Dimensional Ising Model*

Now let's look at a system that does exhibit a phase transition. The first step in the RG theory is to sum over a subset of all the spins in the system. One possible choice of the subset is illustrated in Fig. 5.10. In this figure, the remaining circles represent those spins that have not yet been summed. Notice that the remaining spins form a lattice which is simple square (though rotated by 45°) just like the original lattice.

To carry out the summation over half the spins, we partition the summand in the canonical partition function, Q, so that every other spin appears in only one Boltzmann factor:

$$Q = \sum_{s_1, s_2, \ldots} \cdots \exp\left[Ks_5(s_1 + s_2 + s_3 + s_4)\right]$$
$$\times \exp\left[Ks_6(s_2 + s_3 + s_7 + s_8)\right] \cdots.$$

---

* The development in this section follows closely H. J. Maris and L. J. Kadanoff, *Am. J. Phys.* **46**, 652 (1978).

Thus, by performing the sums over every other spin we get

$$Q = \sum_{\{\text{remaining } s_i\text{'s}\}} \cdots \{\exp\left[K(s_1 + s_2 + s_3 + s_4)\right]$$
$$+ \exp\left[-K(s_1 + s_2 + s_3 + s_4)\right]\}$$
$$\times \{\exp\left[K(s_2 + s_3 + s_7 + s_8)\right]$$
$$+ \exp\left[-K(s_2 + s_3 + s_7 + s_8)\right]\} \cdots.$$

As with the one-dimensional Ising model, we now would like to find a Kadanoff transformation that turns this partially summed partition function into a form that looks just like the original unsummed form. This is not quite possible. Instead, the new partition function is one appropriate to a system with a similar but more general Hamiltonian. To see why, we might try to write

$$\exp\left[K(s_1 + s_2 + s_3 + s_4)\right] + \exp\left[-K(s_1 + s_2 + s_3 + s_4)\right]$$
$$\underset{\boxed{?}}{=} f(K) \exp\left[K'(s_1 s_2 + s_1 s_4 + s_2 s_3 + s_3 s_4)\right]$$

and require that this equation hold for all nonequivalent choices of (s_1, s_2, s_3, s_4). There are four possibilities:

$$s_1 = s_2 = s_3 = s_4 = \pm 1,$$
$$s_1 = s_2 = s_3 = -s_4 = \pm 1,$$
$$s_1 = s_2 = -s_3 = -s_4 = \pm 1,$$
$$s_1 = -s_2 = s_3 = -s_4 = \pm 1.$$

But the supposition above has only two degrees of freedom $f(K)$ and $K'(K)$. Thus, it cannot work.

The simplest possibility that can work is

$$e^{K(s_1 + s_2 + s_3 + s_4)} + e^{-K(s_1 + s_2 + s_3 + s_4)} = f(K) \exp\left[(1/2)K_1\right.$$
$$\times (s_1 s_2 + s_2 s_3 + s_3 s_4 + s_4 s_1)$$
$$\left. + K_2(s_1 s_3 + s_2 s_4) + K_3 s_1 s_2 s_3 s_4\right]. \quad \text{(a)}$$

Inserting the four possibilities for (s_1, s_2, s_3, s_4), we obtain

$$e^{4K} + e^{-4K} = f(K) \exp\left(2K_1 + 2K_2 + K_3\right),$$
$$e^{2K} + e^{-2K} = f(K)e^{-K_3},$$
$$2 = f(K) \exp\left(-2K_2 + K_3\right),$$
$$2 = f(K) \exp\left(-2K_1 + 2K_2 + K_3\right).$$

The solutions exist. They are

$$K_1 = \tfrac{1}{4} \ln \cosh (4K),$$
$$K_2 = \tfrac{1}{8} \ln \cosh (4K),$$
$$K_3 = \tfrac{1}{8} \ln \cosh (4K) - \tfrac{1}{2} \ln \cosh (2K),$$

and

$$f(K) = 2[\cosh (2K)]^{1/2}[\cosh (4K)]^{1/8}.$$

Exercise 5.14 Derive these equations.

Combining Eq. (a) with the partially summed partition function yields

$$Q(K, N) = [f(K)]^{N/2} \sum_{\text{remaining } s_i\text{'s}} \cdots \{\exp [(K_1/2)(s_1 s_2 + s_2 s_3$$
$$+ s_4 s_3 + s_4 s_1) + K_2(s_1 s_3 + s_2 s_4) + K_3 s_1 s_2 s_3 s_4]\}$$
$$\times \{\exp [(K_1/2)(s_2 s_3 + s_3 s_8 + s_7 s_8 + s_7 s_2)$$
$$+ K_2(s_2 s_8 + s_7 s_3) + K_3 s_2 s_7 s_8 s_3]\} \cdots .$$

Notice that every nearest-neighbor pair is repeated exactly twice. For example, $s_2 s_3$ appears in the Boltzmann factor arising from the summing over s_5, and in the one arising from the sum over s_6. However, the next nearest neighbors (e.g., $s_1 s_3$ and $s_2 s_4$) appear only once each; and the sets of four spins around in square (e.g., $s_1 s_2 s_3 s_4$) appear only once each. Thus,

$$Q(K, N) = \sum_{N \text{ spins}} \exp \left[K \sum_{ij}{}' s_i s_j \right]$$
$$= [f(K)]^{N/2} \sum_{N/2 \text{ spins}} \exp \left[K_1 \sum_{ij}{}' s_i s_j \right.$$
$$\left. + K_2 \sum_{lm}{}'' s_l s_m + K_3 \sum_{pqrt}{}''' s_p s_q s_r s_t \right],$$

where the double primed sum is over all *next* nearest-neighbor spins (in the lattice with $N/2$ spins), and the triple primed sum is over all sets of neighboring four spins around in a square.

Notice what has happened. We have removed degrees of freedom, and due to the topology in the two-dimensional system—that is, the high degree of connectivity—the resulting interactions between the remaining degrees of freedom are more complicated than those in the original problem. This occurrence is the usual state of affairs in nontrivial interacting systems. We will see it again when we study liquids. Due to those more complicated interactions, this last equation is not in a form for which an exact RG calculation can be performed. To proceed, we must approximate the summand in the partially summed partition function in such a way that the partially summed quantity resembles the unsummed function. The simplest approximation neglects K_2 and K_3 entirely. That gives

$$Q(K, N) \approx [f(K)]^{N/2} Q(K_1, N/2)$$

with

$$K_1 = \tfrac{1}{4} \ln \cosh (4K).$$

These equations are equivalent to those obtained in the one-dimensional analysis. They predict no phase transition.

To obtain a better result we must account at least for K_2. A simple scheme to do this is a mean-field-like approximation that attempts to incorporate the effects of the non-nearest-neighbor interactions with an altered coupling between nearest neighbors:

$$K_1 \sum_{ij}{}' s_i s_j + K_2 \sum_{lm}{}'' s_l s_m \approx K'(K_1, K_2) \sum_{ij}{}' s_i s_j.$$

This approximation gives

$$Q(K, N) = [f(K)]^{N/2} Q[K'(K_1, K_2), N/2].$$

We let $g(K) = N^{-1} \ln Q(K, N)$ stand for the free energy per spin. As a result

$$g(K) = \tfrac{1}{2} \ln f(K) + \tfrac{1}{2} g(K')$$

or

$$g(K') = 2g(K) - \ln \{2[\cosh (2K)]^{1/2}[\cosh (4K)]^{1/8}\}. \qquad \text{(b)}$$

Exercise 5.15 Derive these formulas.

We can estimate K' by considering the energy of the system when all the spins are aligned. Since there are N nearest-neighbor bonds in a two-dimensional cubic lattice with $N/2$ spins, and there are N next-nearest-neighbor bonds, too,

$$K_1 \sum_{ij}{}' s_i s_j = NK_1, \qquad K_2 \sum_{lm}{}'' s_l s_m = NK_2$$

when all the spins are aligned. As a result, we estimate

$$K' \approx K_1 + K_2$$

or from the equations for $K_1(K)$ and $K_2(K)$,

$$K' = \tfrac{3}{8} \ln \cosh (4K). \qquad \text{(c)}$$

This equation (c) has a nontrivial fixed point! That is, a finite K_c exists for which

$$K_c = \tfrac{3}{8} \ln \cosh (4K_c).$$

Indeed,

$$K_c = 0.50698.$$

Equations (c) and (b) are RG equations that can be solved iteratively to predict the thermodynamic properties of the two-dimensional Ising model. The flow pattern breaks into two parts:

$K = 0$ K_c $K = \infty$

Exercise 5.16 Show that for $K < K_c$, Eq. (c) yields $K' < K$. Similarly, show that for $K > K_c$, Eq. (c) yields $K' > K$.

Since the iterations move away from K_c, this nontrivial fixed point is called an *unstable* fixed point. The trivial fixed points at 0 and ∞, however, are called *stable* fixed points.

To implement the RG equations (b) and (c) it is useful to have their inverses:

$$K = \tfrac{1}{4} \cosh^{-1} (e^{8K'/3}) \qquad\qquad (c')$$

and

$$g(K) = \tfrac{1}{2}g(K') + \tfrac{1}{2} \ln \{2e^{2K'/3}[\cosh(4K'/3)]^{1/4}\}. \qquad (b')$$

Exercise 5.17 Derive these equations.

From Taylor expansions near $K = K_c$, one finds that the heat capacity,

$$C = \frac{d^2}{dK^2} g(K),$$

diverges as $K \to K_c$ according to the power law

$$C \propto |T - T_c|^{-\alpha},$$

where $T = (J/k_B K)$ and

$$\alpha = 2 - \ln 2/\ln (dK'/dK|_{K=K_c})$$
$$= 0.131.$$

Exercise 5.18 Verify this result.

Thus, we associate the fixed point K_c wth a phase transition. The critical temperature is given by

$$\frac{J}{k_B T_c} = 0.50698$$

which is close to the exact value

$$\frac{J}{k_B T_c} = 0.44069$$

obtained from Onsager's solution. The RG prediction of a weakly divergent heat capacity is in qualitative accord with Onsager's result

$$C \propto -\ln |T - T_c|.$$

This application shows that even with crude approximations, the RG theory provides a powerful way of looking at many-body problems. Before ending, however, let us summarize a few of the observations made in this section. One is that the connectivity or topology that produces cooperativity and a phase transition also leads to more and more complicated interactions as one integrates out more and more degrees of freedom. For example, consider again the square lattice pictured at the beginning of this section. "Integrating out" spin 5 means that we have Boltzmann sampled the possible fluctuations of that degree of freedom. Since spin 5 is coupled directly to spins $1, 2, 3,$ and 4, the Boltzmann weighted sum over the fluctuations of spin 5 (i.e., its configurational states) depends upon the particular state of each of these other four spins. Hence, for example, spin 4 "feels" the state of spin 1 through the fluctuations in spin 5. In the lattice that remains after integrating out the first $N/2$ spins, 1 and 4 are not nearest neighbors. Neither are 1 and 3, yet they too are manifestly coupled in the second stage of the RG procedure. This is the origin of the complicated interactions. These new couplings are clearly intrinsic to the connectivity of the model. In the one-dimensional case, this degree of connectivity is absent, one does not generate more complicated interactions by removing degrees of freedom, and no phase transition appears.

Indeed, when we attempt to use the RG procedure and neglect the complicated interactions, we are not able to predict the existence of a phase transition in the two-dimensional model. One way to think about the changes in interactions as one removes degrees of freedom is to imagine a multidimensional interaction parameter space for the coupling constants K_1, K_2, and K_3. In that case, the partition function depends upon all these parameters—that is,

$$Q = Q(K_1, K_2, K_3, \ldots ; N),$$

where ... is used to denote coupling constants for all imaginable interactions (e.g., interactions involving six spins). The partition function of actual interest is $Q(K, 0, 0, \ldots ; N)$, but the first interaction of the RG procedure yields

$$Q(K, 0, 0, \ldots ; N) = [f(K)]^{N/2} Q(K_1, K_2, K_3, 0, \ldots ; N/2).$$

To compute the partition function by an RG procedure, therefore, one must consider transformations of coupling constants in a multidimensional space. It is only by an approximation that we confined the flow to a line in this parameter space.

In closing, we remark that while the theory was originally devised to study second-order phase transitions, the ideas of length scale and Kadanoff transformations, flow in Hamiltonian or coupling parameter space and fixed points extend far beyond this application, and they are finding their way into many diverse areas of physics, chemistry, and engineering.

5.8 Isomorphism between Two-Level Quantum Mechanical System and the Ising Model

One reason for the importance of statistical mechanical techniques, such as the renormalization group method, to various different areas of the physical sciences is due to the isomorphism between quantum theory and classical statistical mechanics. We illustrate the connection here by showing how the statistical behavior of a two-state quantal system is isomorphic to that of the classical Ising model.

We consider the model described in Exercise 3.21. In particular, we imagine a quantum mechanical particle (an electron in a mixed valence compound) fluctuating or resonating between two different localized states. As the particle moves from one position to the other, the deviation from the average of the system's dipole changes sign. This variability of the dipole permits the instantaneous electric field of the surrounding medium to couple to the system. In matrix form, the Hamiltonian and dipole operator for the two-state system are

$$\mathcal{H}_0 = \begin{bmatrix} 0 & -\Delta \\ -\Delta & 0 \end{bmatrix} \quad \text{and} \quad m = \begin{bmatrix} \mu & 0 \\ 0 & -\mu \end{bmatrix}.$$

Exercise 5.19 Show that the eigen energies for this Hamiltonian are $\pm\Delta$ (i.e., the spacing between the two levels is 2Δ), and the eigenvectors are proportional to $(1, \pm 1)$.

The coupling to the electric field, $\mathcal{E}$, leads to the total Hamiltonian

$$\mathcal{H}_0 - \mathcal{E}m.$$

We will assume the surrounding medium is sluggish in comparison to the quantal system. This means that $\mathcal{E}$ is not dynamical and hence, $\mathcal{E}$ is not an operator. For a given $\mathcal{E}$, the partition function of the two level system is

$$Q(\mathcal{E}) = \mathrm{Tr}\, e^{-\beta(\mathcal{H}_0 - m\mathcal{E})},$$

where the trace, Tr, is over the two states of the quantal system.

The next step is the crucial trick. We divide the Boltzmann operator into P identical factors:

$$Q(\mathcal{E}) = \mathrm{Tr}\, [e^{-(\beta/P)(\mathcal{H}_0 - m\mathcal{E})}]^P. \tag{a}$$

For large enough P, we can use the result

$$e^{-(\beta/P)(\mathcal{H}_0 - m\mathcal{E})} = e^{-(\beta/P)\mathcal{H}_0} e^{(\beta/P)m\mathcal{E}}[1 + O(\beta/P)^2]. \tag{b}$$

Exercise 5.20 Expand the exponential operators and verify that the left- and right-hand sides of this equation agree through first order in β/P, and that the deviation at second order involves the commutator $[\mathcal{H}_0, m]$.

Thus, by going to large enough P, we can avoid the mathematical difficulties associated with noncommuting operators. In their place, however, one must account for P separate Boltzmann operators, $\exp[-(\beta/P)(\mathcal{H}_0 - m\mathcal{E})]$. The matrix elements of each of these operators can be analyzed as follows: Let $u = \pm 1$ denote the state of the quantal system. We have

$$\langle u| m |u'\rangle = \delta_{uu'}\mu u \tag{c}$$

and

$$\langle u| \mathcal{H}_0 |u'\rangle = -(1 - \delta_{uu'})\Delta$$
$$= (uu' - 1)\Delta/2. \tag{d}$$

As a result, from (d),

$$\langle u| e^{-\varepsilon\mathcal{H}_0} |u'\rangle = \begin{cases} 1 + O(\varepsilon^2), & u = u' = \pm 1, \\ \varepsilon\Delta + O(\varepsilon^3), & u \neq u' = \pm 1, \end{cases}$$

or

$$\langle u| e^{-\varepsilon\mathcal{H}_0} |u'\rangle = \sqrt{\varepsilon\Delta}\, e^{-uu'\,\ln\sqrt{\varepsilon\Delta}}[1 + O(\varepsilon^2)]. \tag{e}$$

Further, from (c)

$$\langle u| e^{\varepsilon m\mathcal{E}} |u'\rangle = \delta_{uu'}e^{\varepsilon\mu u\mathcal{E}}. \tag{f}$$

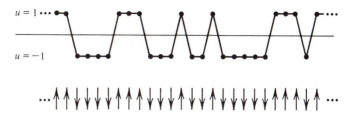

Fig. 5.11. A quantum path (above) and the corresponding configuration of the isomorphic Ising magnet (below).

Hence, combining (b), (e), and (f) yields

$$\langle u | e^{-\varepsilon(\mathcal{H}_0 - m\mathcal{E})} | u' \rangle = \sqrt{\varepsilon\Delta} \exp\left[-uu' \ln\sqrt{\varepsilon\Delta} + \varepsilon\mu\mathcal{E}u\right][1 + O(\varepsilon^2)], \tag{g}$$

where we have introduced the small reciprocal temperature

$$\varepsilon = \beta/P.$$

The trace in (a) can be evaluated employing the usual rule of matrix multiplication—that is,

$$\text{Tr } \mathbf{A}^P = \sum_{u_1, u_2, \ldots, u_P} A_{u_1 u_2} A_{u_2 u_3} \cdots A_{u_P u_1}.$$

Thus, employing (g) we arrive at

$$Q(\mathcal{E}) = \lim_{P \to \infty} \sum_{\substack{u_1, u_2, \ldots, u_P \\ = \pm 1}} (\varepsilon\Delta)^{P/2} \exp\left[\sum_{i=1}^{P} (\kappa u_i u_{i+1} + h u_i)\right],$$

where $\kappa = -\ln\sqrt{\varepsilon\Delta}$ and $h = \varepsilon\mu\mathcal{E}$, and periodic boundary conditions are employed, $u_{P+1} = u_1$. The limit $P \to \infty$ is required to ensure that the terms of the order of ε^2 in (g) can be neglected. This last formula for $Q(\mathcal{E})$ demonstrates the isomorphism since the right-hand side is indeed identical to the partition function for a one-dimensional Ising magnet with a magnetic field.

The method we have used in deriving the isomorphism is the same method introduced by Richard Feynman in the 1940s to derive his path integral formulation of quantum mechanics. In fact, the sum over configurations of the isomorphic Ising magnet is the sum over quantum paths for the two state quantum mechanical system. Adjacent antiparallel spins in the isomorphic magnet correspond to a transition or tunneling event in which the quantal particle moves from one spatial state to the other; that is, the electron in the mixed valence compound resonates between the two nuclei. Figure 5.11 illustrates this isomorphism. The upper portion of the figure pictures a path of the quantal system as it moves between the states with $u = +1$ and $u = -1$. The lower portion pictures the corresponding configuration of the isomorphic Ising magnet. The disorder in the Ising magnet coincides with resonance or quantum tunneling. The ordered isomorphic Ising magnet, on the other hand, corresponds to a spatially localized state in which tunneling does not occur.

In Exercise 5.26 you will consider how the coupling of this two-state model to a slowly *fluctuating* electric field can cause the onset of localization or the quenching of tunneling, and in the isomorphic Ising magnet this localization corresponds to an order-disorder transition. As a preamble to that exercise, let us consider the energy eigenvalues for the problem. For a given value of $\mathscr{E}$, the diagonalization of the 2×2 Hamiltonian yields the following two energy levels (see Exercise 3.21):

$$\pm\sqrt{\Delta^2 + \mu^2 \mathscr{E}^2}.$$

Now imagine that in the absence of the two-state system, the electric field fluctuates slowly and in accord with a Gaussian probability distribution

$$P(\mathscr{E}) \propto \exp(-\beta\mathscr{E}^2/2\sigma),$$

where σ is a parameter that determines the size of the typical electric field fluctuations; indeed, in the absence of the two-state system,

$$\langle \mathscr{E} \rangle = 0$$

and

$$\langle \mathscr{E}^2 \rangle = \sigma/\beta.$$

This information, plus the energy levels for the two-state system coupled to the field, allows us to write down the energies for the full system:

$$E_\pm(\mathscr{E}) = \mathscr{E}^2/2\sigma \pm \sqrt{\Delta^2 + \mu^2 \mathscr{E}^2}.$$

There are two energies, $E_\pm(\mathscr{E})$, since the reversible work to change the field $\mathscr{E}$ depends upon whether the two-state system is in its ground state or excited state. Note that our assumption that the fluctuations in $\mathscr{E}$ are slow is really the assumption that these fluctuations do not cause transitions between the two stationary quantum states. In the usual terminology of quantum mechanics, this assumption is called the adiabatic approximation. We considered this approximation in a different context in Chapter 4 when discussing the Born–Oppenheimer approximation. In Fig. 5.12, the pictures of the corresponding energy surface for this problem show that when the fluctuations in $\mathscr{E}$ are large enough—that is, when $\sigma\mu^2 > \Delta$—the

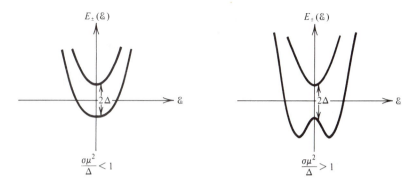

Fig. 5.12. Energies of the two-state system coupled to a field.

configuration with $\mathscr{E} = 0$ is then unstable in the ground state. The stable ground state minima correspond to the situation where the two-state system is coupled to a net electric field. This field breaks the dipolar symmetry of the two-state system leading to a nonzero dipole, and the net polarity corresponds to a localization of the dipole or quenching of its tunneling.

This phase transition-like behavior plays an important role in a variety of condensed phase quantum phenomena including electron transfer processes. After studying Exercise 5.26, the reader may wish to reconsider the phenomena in the following way: The resonating dipole, due to its alternating sign as it passes from one configuration to the other, couples less strongly to the electric field than does the nonresonating localized dipole. The strong interaction is also a favorable one energetically since for large enough σ, the pliable electric field can fluctuate and orient parallel to the direction of the dipole. It is this preference toward a strong and energetically favorable interaction that causes the broken symmetry and localization observed in this model and in the behavior of many molecular systems found in nature. This type of localizaton phenomena is often given the name *self-trapping*.

One last remark before ending this section is a suggestion, a word of caution, and also a word of encouragement: Try to generalize the isomorphism we have derived between the two-state quantum system and the Ising magnet. In particular, consider a quantum system with three (or more) states. You will find that indeed the sampling of quantum paths does map onto a classical statistical mechanics problem. But now, in general, the nearest-neighbor couplings between the multistate "spins" involve complex numbers. This feature means that one must perform Boltzmann weighted sums with alternating signs. In other words, quantum systems with three (or more) states are isomorphic to classical Ising-like systems with negative "probabilities" for certain states. You might consider what feature of the two-state model allows one to avoid this difficulty, and then you might try to invent a method that avoids the problem of negative probabilities for three or more states. If you succeed in the general case, publish!

Additional Exercises

5.21. The canonical partition function for the one-dimensional Ising magnet in an external magnetic field is

$$Q = \sum_{\substack{s_1, s_2, \ldots, s_N \\ = \pm 1}} \exp\left[\sum_{i=1}^{N} (hs_i + Ks_is_{i+1})\right],$$

where $h = \beta\mu H$, $K = \beta J$ and we are using periodic boundary conditions—that is, $s_1 = s_{N+1}$.

(a) Show that Q can be expressed as

$$Q = \text{Tr } \mathbf{q}^N,$$

where $\mathbf{q}$ is the matrix

$$\mathbf{q} = \begin{bmatrix} \exp(-h + K) & \exp(-K) \\ \exp(-K) & \exp(h + K) \end{bmatrix}.$$

[*Hint*: Note that the summand of the argument of the exponential in Q can be rewritten as $h(s_i + s_{i+1})/2 + Ks_is_{i+1}$.]

(b) By noting that the trace of a matrix is independent of representation, show that Q can be expressed as

$$Q = \lambda_+^N + \lambda_-^N,$$

where λ_+ and λ_- are the larger and smaller eigenvalues, respectively, of the matrix $\mathbf{q}$.

(c) Determine these two eigenvalues, and show that in the thermodynamic limit $(N \to \infty)$

$$\frac{\ln Q}{N} = \ln \lambda_+ = K + \ln \{\cosh(h) + [\sinh^2(h) + e^{-4K}]^{1/2}\}.$$

This method for computing a partition function is called the transfer matrix method.

(d) Evaluate the average magnetization and show that the

magnetization vanishes as $h \to 0^+$. [*Hint:* You can determine $\langle s_1 \rangle$ by differentiating $N^{-1} \ln Q$ with respect to h.]

5.22. Consider the one-dimensional Ising model with an external magnetic field. With suitably reduced variables, the canonical partition function is

$$Q(K, h, N) = \sum_{s_1, s_2, \ldots, s_N} \exp\left[h \sum_{i=1}^{N} s_i + K \sum_{i=1}^{N-1} s_i s_{i+1} \right].$$

(a) Show that by summing over all the even spins

$$Q(K, h, N) = [f(K, h)]^{N/2} Q(K', h', N/2),$$

where

$$h' = h + (1/2) \ln[\cosh(2K + h)/\cosh(-2K + h)],$$
$$K' = (1/4) \ln[\cosh(2K + h) \cosh(-2K + h)/\cosh^2(h)],$$

and

$$f(K, h) = 2 \cosh(h)[\cosh(2K + h)$$
$$\times \cosh(-2K + h)/\cosh^2(h)]^{1/4}.$$

(b) Discuss the flow pattern for the renormalization equations, $h' = h'(h, K)$, $K' = K'(K, h)$, in the two-dimensional parameter space (K, h).

(c) Start with the estimate that at $K = 0.01$,

$$g(0.01, h) \approx g(0, h),$$

follow the flow of the RG transformations for several values of h, and estimate $g(1, 1)$ from the RG equations.

5.23. (a) Show how the magnetization per spin in the Ising model corresponds to the density, $\sum_i \langle n_i \rangle / V$, in the lattice gas.

(b) Draw a labeled picture of the coexistence curve in the temperature density plane for the two-dimensional lattice gas.

5.24. Consider a hypothetical system (which forms a closed ring to eliminate end effects) made up of N "partitions." A small section of it is pictured in Fig. 5.13. Each "cell" contains two and only two (hydrogen) atoms—one at the top and one at the bottom. However, each can occupy one of two positions—to the left in the cell (e.g., bottom atom in "cell" jk) or to the right (e.g., bottom atom in "cell" ij or top in "cell" mn). The energies of the possible configurations (a "partition" with its associated atoms) are given by the following rules.

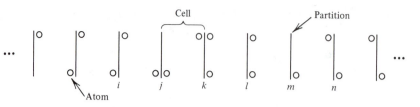

Fig. 5.13. Part of a large system that exhibits a phase transition.

(i) Unless *exactly two atoms* are associated with a partition, the energy of that configuration is (positive) infinite (e.g., $\varepsilon_k = \varepsilon_m = +\infty$).

(ii) If two atoms are on the same side of a partition, then the energy of that configuration is zero (e.g., $\varepsilon_l = 0$).

(iii) If two atoms are on opposite sides of a partition, then the energy of the configuration is ε (e.g., $\varepsilon_i = \varepsilon_n = \varepsilon$).

(a) Using the above rules, what are the energy *levels* possible for the *system* of N partitions and associated atoms?

(b) How many states are present for each level? That is, what is the degeneracy?

(c) Show that the canonical partition function for the *system* is given by one of the following expressions:

$$Q = 2 + 2^N e^{-\beta N \varepsilon},$$
$$Q = 2 + 2N e^{-\beta \varepsilon}, \qquad \beta = 1/k_B T$$
$$Q = 2^N + 2 e^{-\beta N \varepsilon}.$$

(d) Compute the free energy per particle in the thermodynamic limit, and show that the internal energy becomes discontinuous at some temperature, say T_0. (*Hint:* At what value of α is α^m discontinuous for m very large?)

(e) Express T_0 in terms of ε and fundamental constants.

[The problem you have solved represents a (ferroelectric) *phase transition* in the KH_2PO_4 crystal. See *Am. J. Phys.* **36** (1968), 1114. The "partitions" above represent the PO_4 groups.]

5.25. Derive the behavior of the heat capacity at zero field for the heat capacity as predicted from the mean field theory discussed in the text. Determine the precise values just below and above the critical temperature.

5.26. In this exercise, consider the Ising magnet that is isomorphic to the two-state quantal system. Suppose that the electric field, $\mathscr{E}$, is a random fluctuating field with a Gaussian probability distribution,

$$P(\mathscr{E}) \propto \exp[-\beta\mathscr{E}^2/2\sigma].$$

(a) Show that on integrating over $\mathscr{E}$, one obtains the partition function

$$Q = \int_{-\infty}^{\infty} d\mathscr{E}\, e^{-\beta\mathscr{E}^2/2\sigma} Q(\mathscr{E})$$

$$= \sqrt{2\pi\sigma/\beta} \lim_{P\to\infty} \left\{ (\varepsilon\Delta)^{P/2} \sum_{\{u_i\}} \exp\left[\sum_{i=1}^{P} \kappa u_i u_{i+1} \right.\right.$$

$$\left.\left. + (\beta\mu^2\sigma/2P^2) \sum_{i,j=1}^{P} u_i u_j \right] \right\},$$

which is the partition function for a one-dimensional Ising magnet with *long ranged* interactions. Here, 2Δ is the energy spacing of the unperturbed two-state system, $\varepsilon = \beta/P$, and $\kappa = -\ln\sqrt{\varepsilon\Delta}$. See Sec. 5.8.

(b) The long ranged interactions generated by integrating out fluctuations in the electric field can induce a transition in which tunneling is quenched; that is, the fluctuations in the environment cause a spatial localization of the quantal system. Demonstrate that this transition does occur by first evaluating $Q(\mathscr{E})$ and then showing that the Gaussian weighted integral over $\mathscr{E}$ that gives Q is a non-analytic function of σ when $\beta\to\infty$. Identify the critical value of σ at which the transition occurs by considering where $\langle(\delta m)^2\rangle$ diverges. [*Hint*: The square fluctuation is a second derivative of $\ln Q$.] Note that at non-zero temperatures, $\beta\Delta$ is finite; the isomorphic Ising magnet is effectively a finite system in that case, and no localization transition occurs.

5.27. Consider a one-dimensional Ising magnet in a fluctuating magnetic field, h. The particular model to be considered has the partition function

$$Q(\beta, N) = \int_{-\infty}^{\infty} dh \sum_{\substack{s_1,\ldots,s_N \\ =\pm 1}}$$

$$\times \exp\left\{ -\beta N h^2/2\sigma + \sum_{i=1}^{N} [\beta h s_i + \beta J s_i s_{i+1}] \right\}$$

with $s_{N+1} = s_1$. Note that when this system is large (i.e., $N\to\infty$), the typical size of h is very small. Nevertheless, the presence of this small fluctuating field leads to an order-disorder transition for this one-dimensional system.

(a) Integrate out the fluctuating spins with h held fixed and thereby determine the reversible work function for h, $\bar{A}(h;\beta, N)$.
(b) Show that below some temperature (i.e., above some value of β), the free energy $\bar{A}(h;\beta, N)$ becomes a bistable function of h.
(c) Derive an equation for the critical temperature below which this system exhibits the phenomenon of symmetry breaking.

Bibliography

There are several excellent reviews of the modern theory of phase transitions. For example:

B. Widom in *Fundamental Problems in Statistical Mechanics, Vol. III*, ed. by E. D. G. Cohen (North-Holland, 1975) pp. 1–45.

M. E. Fisher in *Critical Phenomena, Lecture Notes in Physics, Vol. 186*, ed. by F. J. W. Hahne (Springer-Verlag, N.Y., 1983).

K. G. Wilson, *Rev. Mod. Phys.* **55,** 583 (1983).

The last of these is based on Wilson's Nobel Prize award address. It contains a brief review of the renormalization group theory, and also Wilson's description of the history of this theory.

The late Shang-Keng Ma's text contains many useful ideas on Ising models, on boundary and surface effects in phase transformations, and on mean field theories:

S. K. Ma, *Statistical Mechanics* (World Scientific, Philadelphia, 1985).

Self-similarity or invariance to length scale transformations is central to renormalization group theory. It is also central to the theory of "fractals"— irregular geometrical structures (such as coastlines). A popularized description of fractals is given in Mandelbrot's book:

B. B. Mandelbrot, *The Fractal Geometry of Nature* (Freeman, San Francisco, 1982).

A general text on the subject of phase transitions is Stanley's soon to be revised book:

H. E. Stanley, *Introduction to Phase Transitions and Critical Phenomena* (Oxford University Press, London, 1972).

In the last section of this chapter, we introduced path integrals and the connection between quantum theory in Euclidean (imaginary) time and statistical mechanics. The standard text on the subject is

R. P. Feynman and A. R. Hibbs, *Path Integrals and Quantum Mechanics* (McGraw Hill, N.Y., 1965).

CHAPTER 6

Monte Carlo Method in Statistical Mechanics

With the advent and wide availability of powerful computers, the methodology of computer simulations has become a ubiquitous tool in the study of many-body systems. The basic idea in these methods is that with a computer, one may follow explicitly the trajectory of a system involving 10^2 or 10^3 or even 10^4 degrees of freedom. If the system is appropriately constructed—that is, if physically meaningful boundary conditions and interparticle interactions are employed—the trajectory will serve to simulate the behavior of a real assembly of particles, and the statistical analysis of the trajectory will determine meaningful predictions for properties of the assembly.

The importance of these methods is that, in principle, they provide exact results for the Hamiltonian under investigation. Thus, simulations provide indispensable bench marks for approximate treatments of non-trivial systems composed of interacting particles. Often, the simulations themselves are so efficient that they can be readily performed for all circumstances of interest, and there is no need for recourse to approximate and computationally simpler approaches. There are, however, important limitations. The finite capacity of computers (both in memory and in time) implies that one may consider only a finite number of particles, and one may follow trajectories of only finite length. The latter restriction places bounds on the quality of the statistics one can acquire, and the former inhibits the study of large length scale fluctuations. These issues will become clearer as we become more specific with particular illustrations.

There are two general classes of simulations. One is called the *molecular dynamics* method. Here, one considers a classical dynamical model for atoms and molecules, and the trajectory is formed by integrating Newton's equations of motion. The procedure provides dynamical information as well as equilibrium statistical properties. The other class is called the *Monte Carlo* method. This procedure is more generally applicable than molecular dynamics in that it can be used to study quantum systems and lattice models as well as classical assemblies of molecules. The Monte Carlo method does not, however, provide a straightforward method of obtaining time-dependent dynamical information. In this chapter, our discussion of Monte Carlo will focus on lattice models, the Ising magnet, and the lattice gas, systems for which we already have some experience. The illustrations are easily generalized to more complex problems.

Several computer codes are presented in this and later chapters. They are all written in BASIC, and run on microcomputers. The student should experiment with these programs as well as the generalizations outlined in the Exercises. The experimentation is indispensable for gaining a qualitative understanding of the power and limitations of simulation calculations. In all cases, the models analyzed in this chapter are models for which much is known from exact analytical results. These exact results provide guidance in the experimentation, and it is always useful to test simulation algorithms against such results before venturing into the less well known. In Chapter 7, we do step into that territory by presenting a Monte Carlo program and calculation for a model of a liquid (albeit in two dimensions).

6.1 Trajectories

A trajectory is a chronological sequence of configurations for a system. For example, the configuration of a lattice gas or Ising magnet is the list of spin variables $s_1, s_2, \ldots, s_N$. Let $v = (s_1, \ldots, s_N)$ be an abbreviation for a point in this N-dimensional configuration space. Now imagine a path through this space. Let $v(t)$ denote the list $s_1, s_2, \ldots, s_N$ at the tth step in this path. The path function $v(t)$ is then a trajectory. Schematically, we might imagine the graph in Fig. 6.1 for the first eight steps in a trajectory. The letters a, b, c, and d refer to different configurations. For example, perhaps $a = (1, 1, -1, 1, \ldots)$, $b = (1, -1, -1, 1, \ldots)$, $c = (-1, -1, -1, 1, \ldots)$, and $d = (1, -1, 1, -1, \ldots)$.

Configurational properties change as the trajectory progresses, and the average of a property $G_v = G(s_1, s_2, \ldots, s_N)$ over the configurations visited during a trajectory with T steps is

$$\langle G \rangle_T = \frac{1}{T} \sum_{t=1}^{T} G_{v(t)}.$$

Fig. 6.1. Trajectory.

In Monte Carlo calculations, one usually employs trajectories for which thermally equilibrated averages, $\langle G \rangle$, are given by

$$\langle G \rangle = \lim_{T \to \infty} \langle G \rangle_T.$$

That is, the trajectories are ergodic, and constructed such that the Boltzmann distribution law is in accord with the relative frequencies with which different configurations are visited. In practice, trajectories are carried out for only a finite time duration, and the average over configurations will provide only an estimate of $\langle G \rangle$.

We can visualize the limited statistical accuracy of a finite time average by considering the cumulative average as a function of T. See Fig. 6.2. The size of the oscillations about $\langle G \rangle$ coincides with the statistical uncertainty in the average. This point is understood by dividing a long trajectory into several shorter successive trajectories. If the short trajectories are not too short, then the averages over each of these subtrajectories can be viewed as statistically independent observations, and a comparison between them can be used to estimate standard deviations. Schematically, this idea is illustrated with the picture of the time line divided into increments of length L:

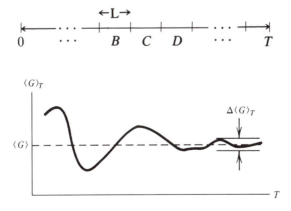

Fig. 6.2. Cumulative average.

Let $\langle G \rangle^{(B)}$ denote the average of $G_{\nu(t)}$ over the Bth increment. Clearly,

$$\langle G \rangle_T = (L/T) \sum_{I=A,B,\ldots} \langle G \rangle^{(I)}.$$

Further, a calculation of the standard deviation gives an estimate of the statistical uncertainty

$$\Delta\langle G \rangle_T = \left[(L/T)^2 \sum_I (\langle G \rangle^{(I)} - \langle G \rangle_T)^2 \right]^{1/2}.$$

As $T \to \infty$, this statistical error tends to zero, vanishing as $T^{-1/2}$.

Exercise 6.1 Show that $\Delta\langle G \rangle_T$ given by this formula is indeed an estimate of the size of the oscillations shown in the figure depicting the cumulative averages $\langle G \rangle_T$.

Unfortunately, straightforward applications of this type of analysis are not always reliable indicators of statistical errors. Problems arise when the system may be sluggish in the sense that trajectories move slowly through available configuration space or the trajectories are trapped in a subspace of the accessible configurations. This sluggish behavior may arise from the physical nature of the system or it may be a consequence of algorithm by which the trajectory is performed. In the quasi-ergodic case where the trajectory samples only a subset of available configurations, one can imagine a large energy barrier that confines the system to this region. When the system is sluggish, the time or number of steps, L, between statistically independent sections of the trajectory becomes very large. For such a case, one might be tricked into thinking that excellent statistics have been obtained, when in fact, the system has progressed through only a subset of the allowed fluctuations.

A test of the statistical independence can be obtained by computing correlation functions of the type

$$(L/T) \sum_I (\langle G \rangle^{(I)} - \langle G_T \rangle)(\langle G \rangle^{(I+1)} - \langle G \rangle_T)$$

and comparing the square root of this quantity with $\langle G \rangle_T$. However, even when the ratio is very small for a particular property G, it may still be that some other (slow) property possesses significant correlations between successive increments. These thoughts should always be a source of concern when employing simulations. A certain degree of experimentation is always required to indicate whether the statistical results are reliable. And while the indications can be convincing, they can never be definitive.

Exercise 6.2 Describe a model system for which you can conceive of circumstances where Newtonian trajectories will be quasi-ergodic. One example might be a system of spherical particles jammed into a fixed volume and into a metastable lattice configuration. Note that if one performed trajectories where the volume of the container could fluctuate and deform, the trajectory could eventually move to stable lattice configurations.

6.2 A Monte Carlo Trajectory

Now we consider a rule or algorithm for carrying out a trajectory. The scheme does not correspond to a true dynamics. However, it is a method for performing a random walk through configuration space. Further, we will show that in the absence of quasi-ergodic problems, averages over these statistical trajectories do correspond to equilibrium ensemble averages. The method is called *Monte Carlo*. (The origin of the name has to do with its exploitation of random number sequences—the type of sequence encountered when playing roulette in that European city.)

For any large non-trivial system, the total number of possible configurations is an astronomically large number, and straightforward sampling of all of these configurations is impractical. A similar situation is encountered by pollsters trying to gauge preferences of large groups of people. In that case, the researcher devises an efficient method by which estimates are obtained from a relatively small but representative fraction of the total population. Monte Carlo calculations are analogous. A sample of representative states or configurations is created by performing a random walk through configuration space in which the relative frequency of visitations is consistent with the equilibrium ensemble distribution. For a small two-dimensional Ising magnet with only $20 \times 20 = 400$ spins, the total number of configurations is $2^{400} > 10^{100}$. Yet Monte Carlo procedures routinely treat such systems successfully sampling only 10^6 configurations. The reason is that Monte Carlo schemes are devised in such a way that the trajectory probes primarily those states that are statistically most important. The vast majority of the 2^{400} states are of such high energy that they have negligible weight in the Boltzmann distribution, and need not be sampled. The words *importance sampling* are used to describe schemes that are biased to probe the statistically important regions of configuration space.

Because the Monte Carlo procedure is not truly dynamical, there exists a great deal of flexibility in choosing the particular algorithm by which the random walk is performed. Here, we give one reasonable method for an Ising magnet. We begin by setting up our system in an initial configuration. Then we pick out at random one of the spins in this assembly. The random choice for the spin is performed with the aid of the *pseudo-random number generator* (an algorithm that generates a long sequence of random numbers uniformly distributed in the interval between 0 and 1).* Since the N spins are labeled perhaps as $s(I)$ for $1 \leq I \leq N$, we can identify a spin by picking I as the closest integer to a generated random number times N. Other methods could be used too. For example, in a two-dimensional Ising model, convenient bookkeeping might lead us to label spins with a vector $s(I, J)$, where $1 \leq I, J \leq \sqrt{N}$. In that case, we could generate two random numbers, x and y, and ask for the integers, I and J, which are closest to $\sqrt{N} x$ and $\sqrt{N} y$, respectively.

Having picked out a spin at random, we next consider the new configuration v', generated from the old configuration, v, by flipping the randomly identified spin. For example, $v \to v'$ could correspond to

$$(\ldots, 1, -1, 1, 1, \ldots) \longrightarrow (\ldots, 1, 1, 1, 1, \ldots).$$

↑   ↑
randomly identified   randomly identified
spin   spin

The change in configuration changes the energy of the system by the amount

$$\Delta E_{vv'} = E_{v'} - E_v.$$

This energy difference governs the relative probability of configurations through the Boltzmann distribution, and we can build this probability into the Monte Carlo trajectory by a criterion for accepting and rejecting moves to new configurations.

In particular, if the energy change $\Delta E_{vv'}$ is negative or zero, we accept the move. If, however, $\Delta E_{vv'}$ is positive, we draw a random number x between 0 and 1, and accept the move only if we have $\exp(-\beta \Delta E_{vv'}) \geq x$. Otherwise the move to a new configuration in the next step is rejected. In other words, with

$$v(t) = v,$$

we have

$$v(t + 1) = v' \quad \text{when } \Delta E_{vv'} \leq 0,$$

---

* Pseudo-random number generators are commonly available with digital computers as standard software and/or hardware.

and

$$v(t+1) = \begin{cases} v', & \exp(-\beta \Delta E_{vv'}) \geqslant x, \\ v, & \exp(-\beta \Delta E_{vv'}) < x, \end{cases}$$

when $\Delta E_{vv'} > 0$. We now repeat this procedure for making a step millions of times thus forming a long trajectory through configuration space.

This algorithm for a trajectory is often called Metropolis Monte Carlo in recognition of the scientists N. Metropolis, A. Rosenbluth, M. Rosenbluth, A. Teller, and E. Teller who first published this type of calculation in 1953. It is easy to argue that the procedure does indeed provide a method for performing equilibrium averages. To see this point we can phrase the procedure in terms of a transition probability matrix. That is, let

$w_{vv'}$ = probability per unit time that if the system is in
state v, it will make a transition to state v'.

If we follow a first-order kinetics associated with this rate or transition matrix, we have the equation

$$\dot{p}_v = \sum_{v'} [-w_{vv'} p_v + w_{v'v} p_{v'}],$$

where p_v gives the probability that the trajectory resides at a given time in state v. (Kinetic equations of this form are often called *master equations*.) At equilibrium in the canonical ensemble $\dot{p}_v = 0$ and

$$(p_{v'}/p_v) = \exp(-\beta \Delta E_{vv'}).$$

These equations provide the condition of detailed balance

$$(w_{vv'}/w_{v'v}) = (p_{v'}/p_v) = \exp(-\beta \Delta E_{vv'}).$$

Provided a trajectory algorithm obeys this condition, the statistics acquired from the trajectory will coincide to those of the equilibrium canonical ensemble. In the Metropolis algorithm,

$$w_{vv'} \propto \begin{cases} 1, & \Delta E_{vv'} \leqslant 0, \\ \exp(-\beta \Delta E_{vv'}), & \Delta E_{vv'} \geqslant 0. \end{cases}$$

That is, there is unit probability of accepting a move when the move causes an energy decrease, and when the energy change is positive, the move is accepted with an exponential probability. (Note: The comparison of the uniform random number distribution with the Boltzmann factor generates this exponential probability distribution.) Given this formula for $w_{vv'}$, it is clear that the Metropolis Monte Carlo generates a trajectory that samples configurations in accord with the canonical Boltzmann distribution of configurations.

Exercise 6.3 Construct and describe an algorithm which employs a random number generator and samples states of the Ising magnet consistent with the distribution

$$P_v \propto \exp(-\beta E_v + \xi M_v),$$

where M_v is the magnetization of the Ising model for spin configuration v.

Exercise 6.4 Describe a Monte Carlo algorithm that samples canonical equilibrium for a model with

$$E_v = -(J/100) \sum_{ij}' s_i s_j, \qquad v = (s_1, s_2, \ldots, s_N),$$

where the primed sum is over nearest neighbors on a square lattice and the spins, s_i, can take on several integer values:

$$-10 \leqslant s_i \leqslant 10.$$

Demonstrate that the algorithm does obey detailed balance. Note that when constructing this algorithm, you must specify step sizes for the attempted moves. Consider the possibility of attempting steps for which $\Delta s_i = s_i(t+1) - s_i(t)$ is larger in magnitude than 1. Will the average acceptance of attempted moves depend upon the average size of Δs_i?

A Monte Carlo written in BASIC for the two-dimensional Ising magnet in a square lattice with $20 \times 20 = 400$ spins appears at the end of this chapter. It utilizes the random number generator provided by the IBM PC computer.

The pictures in Fig. 6.3 show the evolution of the Monte Carlo trajectory for a few thousand steps. To interpret the pictures, note that periodic boundary conditions are employed. At temperatures below critical $(k_B T_c/J \approx 2)$, phase separation remains stable. But except at very low temperatures, the interface clearly fluctuates in an amoeboid fashion. At high temperatures, the trajectory quickly moves to a highly disordered state. Compare that disorder to the type of fluctuations observed near the critical temperature. Note that the patterns formed by the fluctuations at $T \approx T_c$ are very different than the patterns formed when T is much larger than T_c. The study of correlations between fluctuations at different points in space provides one method for quantifying this pattern recognition. Incidentally, a few thousand moves are not nearly enough to acquire reasonable

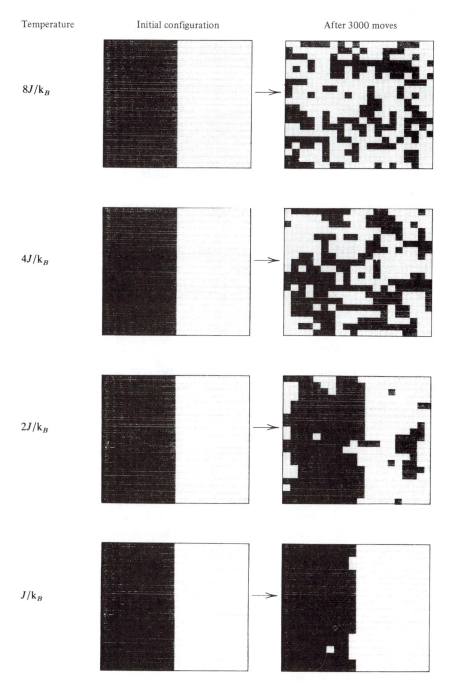

Fig. 6.3. Configurations in a Monte Carlo trajectory of a 20 × 20 Ising model.

statistics from a Monte Carlo trajectory. Further, 400 spins is far too small a system if one is really interested in acquiring quantitative results for critical fluctuations. Nevertheless, the system is simple enough to be examined with a microcomputer and serves as a very instructive pedagogical tool.

Exercise 6.5 Run this Monte Carlo code on a computer. Evaluate the average of the spin (for every hundredth step in the trajectory, the code lists the value of the fraction of up spins). Demonstrate that the average spin is zero for temperatures higher than critical, and try to exhibit the phenomenon of spontaneous symmetry breaking by showing that the average spin is not zero when the temperature is below critical. Explore the dependence of your observations on initial configurations of the magnet (the code provides you with three immediate choices for initial configurations; with slight alteration you can enlarge on this variety), and on the time duration of the sampling.

6.3 Non-Boltzmann Sampling

Suppose it is convenient to generate a Monte Carlo trajectory for a system with energy $E_v^{(0)}$, but you are actually interested in the averages for a system with a different energetics

$$E_v = E_v^{(0)} + \Delta E_v.$$

For example, suppose we wanted to analyze a generalized Ising model with couplings between certain non-nearest neighbors as well as nearest neighbors. The convenient trajectory might then correspond to that for the simple Ising magnet, and ΔE_v would be the sum of all non-nearest-neighbor interactions. How might we proceed to evaluate the averages of interest from the convenient trajectories? The answer is obtained from a factorization of the Boltzmann factor:

$$e^{-\beta E_v} = e^{-\beta E_v^{(0)}} e^{-\beta \Delta E_v}.$$

With this simple property we have

$$Q = \sum_v e^{-\beta E_v} = Q_0 \sum_v e^{-\beta E_v^{(0)}} e^{-\beta \Delta E_v}/Q_0$$

$$= Q_0 \langle e^{-\beta \Delta E_v} \rangle_0,$$

where $\langle \cdots \rangle_0$ indicates the canonical ensemble average taken with energetics $E_v^{(0)}$. We have already encountered this result when considering the method of thermodynamic perturbation theory in Sec. 5.5. A similar result due to the factorization of the Boltzmann factor is

$$\langle G \rangle = Q^{-1} \sum_v G_v e^{-\beta E_v}$$
$$= (Q_0/Q) \langle G_v e^{-\beta \Delta E_v} \rangle_0$$
$$= \langle G_v e^{-\beta \Delta E_v} \rangle_0 / \langle e^{-\beta \Delta E_v} \rangle_0.$$

These formulas provide a basis for Monte Carlo procedures called *non-Boltzmann sampling* and *umbrella sampling*. In particular, since a Metropolis Monte Carlo trajectory with energy $E_v^{(0)}$ can be used to compute the averages denoted by $\langle \cdots \rangle_0$, we can use the factorization formulas to compute $\langle G \rangle$ and (Q/Q_0) even though the sampling employing $E_v^{(0)}$ is not consistent with the Boltzmann distribution, $\exp(-\beta E_v)$.

The simplest idea one might have in this regard is to compute the total free energy from $\ln(Q/Q_0)$ by taking $E_v^{(0)} = 0$. That idea is usually not a good one, however, since the trajectory will be unbiased and much of the trajectory can be wasted exploring regions of configuration space that are inaccessible when the energy is E_v. The point of Monte Carlo is to avoid such inefficient exploration. Non-Boltzmann sampling is a powerful tool when the reference or unperturbed energy, $E_v^{(0)}$, creates a trajectory that is close to that for E_v.

Exercise 6.6 Consider an Ising magnet at a temperature T and another at a different temperature T'. Show that the difference in free energy per unit temperature between the two can be computed by averaging

$$\exp\left[-E_v \left(\frac{1}{k_B T'} - \frac{1}{k_B T} \right) \right]$$

over a Monte Carlo trajectory for the system at one or the other temperature.

Non-Boltzmann sampling can also be useful in removing the bottlenecks that create quasi-ergodic problems and in focusing attention on rare events. Consider the former first. Suppose a large activation barrier separates one region of configuration space from another, and suppose this barrier can be identified and located at a configuration or set of configurations. We can then form a reference system in which the barrier has been removed; that is, we take

$$E_v^{(0)} = E_v - V_v,$$

where V_v is large in the region where E_v has the barrier, and it is zero otherwise. Trajectories based upon the energy $E_v^{(0)}$ will indeed waste time in the barrier region, but they will also pass from one side of the barrier to the other thus providing a scheme for solving problems of quasi-ergodicity.

The other situation we referred to occurs when we are interested in a relatively rare event. For example, suppose we wanted to analyze the behavior of surrounding spins given that an $n \times n$ block were all perfectly aligned. While such blocks do form spontaneously, and their existence might catalyze something of interest, the natural occurrence of this perfectly aligned $n \times n$ block might be very infrequent during the course of the Monte Carlo trajectory.

Exercise 6.7 Consider an analogous problem encountered in solution theory. Two solutes are dissolved in a fluid of 400 solvent molecules to simulate a solution at low solute concentrations. Suppose the solutes interact strongly with each other and we are interested in studying these interactions as they are mediated by the solvent. Then we are interested in only those configurations for which the solutes are close to each other. Use a lattice model and estimate the fraction of accessible configurations in the total system for which the solutes will be close together. (The fraction is a terribly small number.)

How are we to obtain meaningful statistics for these rare events without wasting time with irrelevant though accessible configurations? The answer is as follows:

Perform a non-Boltzmann Monte Carlo trajectory with an energy like:

$$E_v^{(0)} = E_v + W_v,$$

where W_v is zero for the interesting class of configurations, and for all others, W_v is very large. The energy W_v is then called an umbrella potential. It biases the Monte Carlo trajectory to sample only the rare configurations of interest. The non-Boltzmann sampling performed in this way is called *umbrella sampling*.

To illustrate this methodology, we consider in the remainder of this section the computation of the free energy function, $\bar{A}(M)$ (see

Sec. 5.3). This function is defined for the Ising magnet by

$$\exp[-\beta\bar{A}(M)] = \sum_{\nu} \Delta\left(M - \mu \sum_{i=1}^{N} s_i\right) \exp(-\beta E_{\nu}),$$

where $\Delta(x)$ is the Kronecker delta (it is 1 when $x = 0$, and zero otherwise). That is, $\exp[-\beta\bar{A}(M)]$ is the Boltzmann weighted sum over those states for which the net magnetization is M. Clearly,

$$\exp[-\beta\bar{A}(M)] \propto P(M) = \left\langle \Delta\left(M - \mu \sum_{i=1}^{N} s_i\right) \right\rangle,$$

where $P(M)$ is the probability for observing the Ising magnet with magnetization M. [In the limit of an infinite system, the integer spacing between different values of M/μ becomes an infinitesimal spacing in comparison with the total range of M/μ, which is between $-N$ and N. In that limit, the Kronecker delta could be replaced by a Dirac delta function, and $P(M)$ would then be a probability *distribution*.]

In a straightforward computation of $\bar{A}(M)$, one would analyze the number of times states of a given magnetization are visited during a Monte Carlo trajectory. The histogram derived from this analysis is proportional to $P(M)$, and its logarithm determines $\bar{A}(M)$. In many circumstances, such a procedure is completely satisfactory. However, if we consider the situation of broken symmetry (i.e., when $T < T_c$), and plan to compute $\bar{A}(M)$ for a wide range of M values, we immediately encounter a serious problem. In particular, for $T < T_c$, $\bar{A}(M)$ is a bistable function of M, and for a large system, the great majority of states have negligible statistical weight in comparison to those for which $M \approx \pm Nm\mu$ (here, $m\mu$ is the spontaneous magnetization per spin). For example, even for the relatively small system of $20 \times 20 = 400$ spins, at $k_B T/J \approx 1$, the surface energy is $\sim 10 k_B T$, and therefore the probability of states with $M = 0$ is about $\exp(-10)$ less than those with broken symmetry. As a result, the visitation of states with $M = 0$ is an infrequent event, and as such, relatively poor statistics will be acquired for these regions of infrequent visitations.

The method of umbrella sampling, however, avoids this difficulty. We chose a set of umbrella or window potentials

$$W_{\nu} = 0, \quad \text{for } M_i - w/2 \leq \mu \sum_{j=1}^{N} s_j \leq M_i + w/2$$

$$= \infty, \quad \text{otherwise.}$$

For each of these potentials—that is, for each M_i—a simulation is performed. Within each window, the analysis of a histogram is

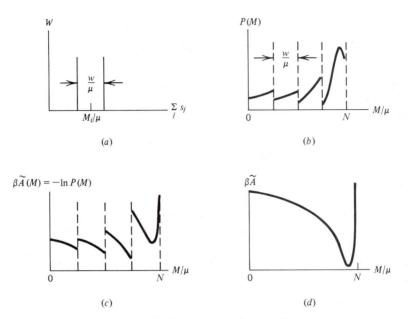

Fig. 6.4. Method of umbrella sampling.

performed to determine the probability for the magnetization in the range between $M_i - w/2$ and $M_i + w/2$. After the full range of magnetization is studied in this way [which requires a minimum of $(N\mu/w)$ separate simulations], the full $P(M)$ and thus $\tilde{A}(M)$ is found by requiring that the latter of these functions is a continuous function from one window to the next. Here, note that with this procedure, $\tilde{A}(M)$ is determined in each window to within an additive constant. The presence of this constant, which must be adjusted from one window to the next, is a consequence of the fact that the histograms determine the probabilities in each window to within a constant of normalization.

Figure 6.4 is a schematic of this procedure. Picture (a) shows one of the window potentials; (b) and (c) depict a set of probabilities and free energies $\tilde{A}(M) = -\beta^{-1} \ln P(M)$, respectively, that would be acquired under ideal circumstances from each window; Picture (d) shows the continuous curve that would be obtained by connecting the $\tilde{A}(M)$ from each region assuming windows have points in common with their neighbors.

One way of thinking about this procedure is that by moving from one window to the next, you are propelling the system reversibly through the relevant states. It is "reversible" because within each window, all states or fluctuations are sampled in accord with the

Boltzmann distribution. Provided the variation of $\bar{A}(M)$ is no more than 1 or $2k_B T$ within each window, one should be able (with a long enough trajectory) to accurately sample the statistics in each window. Let τ denote the required computer time to acquire such statistics within each window. Then, the total computation time to determine $\bar{A}(M)$ is $n\tau$, where n is the number of windows necessary to span the total range of M. Note that*

$$\tau \propto w^2.$$

Therefore, the total computation time required to determine $\bar{A}(M)$ by the method of umbrella sampling is

$$t_{\text{CPU}} \propto nw^2.$$

Now, how much time would it have taken if we did not use this method? As a lower bound, let us assume that $\bar{A}(M)$ does not vary more than a few $k_B T$ over the entire range of M. The size of this range is nw. Therefore,* the time to sample this range is proportional to $(nw)^2 = nt_{\text{CPU}}$. Hence, without the windows, the computation time would be n times longer than that with n windows. The advantage (i.e., lower computation time) of umbrella sampling is, of course, even greater than this when regions of M have relatively high values of $\bar{A}(M)$ and thus relatively low probabilities.

Exercise 6.8 This argument might suggest that the ultimate efficiency would be obtained by choosing extremely narrow windows. Why is this argument incorrect? [*Hint*: Think about the acceptance rate of steps in the Monte Carlo trajectory.]

The BASIC code for the 400 spin Ising magnet has been modified to perform an umbrella sampling calculation of $\bar{A}(M)$. Figure 6.5 presents some representative results obtained in this way.

The calculations employed a window width of 40μ. That is, 10 windows were used between $M = 0$ and $M = 400\mu$. In each window, long trajectories were performed. The algorithm is the same as that for the code presented in Sec. 6.2 except that an additional rejection criteria is employed. In particular, moves are rejected if the total magnetization, $\mu \sum_i s_i$, falls outside of the specified window. Otherwise, the acceptance-rejection criteria are identical to the Metropolis scheme. The number of "passes" referred to in the figure refers to

---

* The argument that establishes this proportionality requires one to know that in a random walk or diffusive processes, such as Monte Carlo, the mean square distance traversed in a time t is proportional to t. We discuss diffusive motion in Chapter 8.

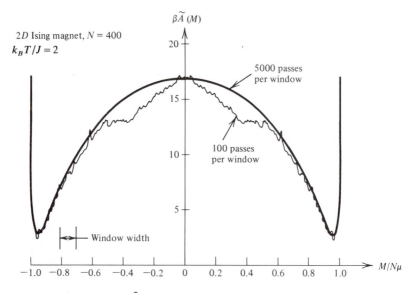

Fig. 6.5. $\tilde{A}(M)$ as calculated by umbrella sampling.

the length of the trajectories performed in each window. A pass denotes $N = 400$ attempted moves. Thus, 100 passes refers to a trajectory of 40,000 steps. The statistics obtained from this trajectory are apparently not nearly as good as those obtained with 5000 passes.

Before leaving this section, it is of interest to consider the qualitative appearance of the free energy, $\tilde{A}(M)$, we have computed. On viewing Fig. 6.5, we see that the calculations do indeed demonstrate the feature of broken symmetry where stable states exist for $M \neq 0$. Further, we see that the barrier is rather flat, and the wells are rather narrow. Can you predict how these features depend upon the size of the system? First, let's think about the stable states. If plotted as a function of $M/N\mu$, the minima of $\beta\tilde{A}(M)$ will become narrower as N increases. The reason is that the width of a well determines for one of the stable states the typical size of the spontaneous fluctuations in the order parameter, and since $\langle (\delta M)^2 \rangle = \partial \langle M \rangle / \partial \beta H$, the root mean square fluctuation of $M/N\mu$ is of the order of $1/\sqrt{N}$.

Now what about the barrier separating the stable states? Here, as N increases, the barrier seen in Fig. 6.5 should get both higher and flatter. The reasoning behind this prediction focuses on the energetics of forming a surface and surface excitations—it is left as a puzzle for the student to think about. (*Hint:* Consider first the energetics of changing a region of an ordered Ising magnet into a region of opposite spin. To do this, a surface must be created, and the magnitude of the net magnetization will be reduced. Next, consider the energetics of the surface excitations required to further reduce M.)

6.4 Quantum Monte Carlo

In Sec. 5.8 we introduced the concept of discretized quantum paths and illustrated how, in this representation, quantum mechanics is isomorphic with the Boltzmann weighted sampling of fluctuations in a discretized classical system with many degrees of freedom. Since such sampling can be done by Monte Carlo, this mapping provides a basis for performing quantum Monte Carlo calculations. This particular approach, treated in this section, is usually given the full name *path integral quantum Monte Carlo* to distinguish it from a host of other Monte Carlo sampling schemes used to generate numerical solutions to Schrödinger's equation. Path integral Monte Carlo is the method of choice for studying quantum systems at non-zero temperatures.

We will keep the discussion at a relatively simple level by considering only one model: the two-state quantum system coupled to a Gaussian fluctuating field. This is the model examined in Sec. 5.8 and in Exercise 5.26. It is a system that can be analyzed analytically, and comparison of the exact analytical treatment with the quantum Monte Carlo procedure serves as a useful illustration of the convergence of Monte Carlo sampling.

According to the analysis of the model presented in Sec. 5.8 and Exercise 5.26, the quantal fluctuations are to be sampled from the distribution

$$W(u_1, \ldots, u_P; \mathscr{E}) \propto \exp[\mathscr{S}(u_1, \ldots, u_P; \mathscr{E})],$$

where

$$\mathscr{S} = -\beta \mathscr{E}^2/2\sigma + (\beta/P)\mu \sum_{i=1}^{P} u_i \mathscr{E}$$

$$- \tfrac{1}{2} \ln(\beta \Delta/P) \sum_{i=1}^{P} u_i u_{i+1}.$$

Here,

$$u_i = \pm 1$$

specifies the state of the quantum system at the ith point on the quantum path, there are P such points, and periodicity requires $u_{P+1} = u_1$. As in Sec. 5.8, the parameters μ and Δ correspond to the magnitude of the dipole and half the tunnel splitting of the two-state system, respectively. The electric field, $\mathscr{E}$, is a continuous variable that fluctuates between $-\infty$ and ∞, but due to the finite value of σ, the values of $\mathscr{E}$ close to $\sqrt{\sigma/\beta}$ in magnitude are quite common.

The quantity $\mathscr{S}$ is often called the *action*. It is a function of the P points on the quantum path $(u_1, \ldots, u_P)$. The discretized representation of the quantum path becomes exact in the continuum limit of an infinite number of points on the path—that is, $P \to \infty$. In that limit, the path changes from a set of P variables, u_1 through u_P, to a function of a continuous variable. The action is then a quantity whose value depends upon a function, and such a quantity is called a *functional*.

A computer code presented at the end of the chapter, written in BASIC to run on an IBM PC, samples the discretized weight function $W(u_1, \ldots, u_P; \mathscr{E})$ by the following procedure: Begin with a given configuration. One of the $P+1$ variables u_1 through u_P and $\mathscr{E}$ is identified at random, and the identified variable is changed. For instance, if the variable is u_i, the change corresponds to $u_i \to -u_i$. If, on the other hand, the identified variable is the field $\mathscr{E}$, then the change is $\mathscr{E} \to \mathscr{E} + \Delta\mathscr{E}$, where the size of $\Delta\mathscr{E}$ is taken at random from a continuous set of numbers generated with the aid of the pseudo-random-number generator.

This change is one of the variables causes a change in the action, $\Delta\mathscr{S}$. For example, if $u_i \to -u_i$, then

$$\Delta\mathscr{S} = \mathscr{S}(u_1, \ldots, -u_i, \ldots; \mathscr{E}) - \mathscr{S}(u_1, \ldots, u_i, \ldots; \mathscr{E}).$$

The Boltzmann factor associated with this change, $\exp(\Delta\mathscr{S})$, is then compared with a random number x taken from the uniform distribution between 0 and 1. If

$$\exp(\Delta\mathscr{S}) > x,$$

the move is accepted. [Note that no explicit comparison is required if $\Delta\mathscr{S} > 0$ since in that case $\exp(\Delta\mathscr{S}) > 1$.] If, however,

$$\exp(\Delta\mathscr{S}) < x,$$

the move is rejected. An accepted move means that the new configuration of the system is the changed configuration. A rejected move means that the new configuration is unchanged from the previous one.

As with the Ising model studied in Sec. 6.2, this standard Metropolis procedure for a Monte Carlo step is repeated over and over again producing a trajectory that samples configuration space according to the statistical weight $W(u_1, \ldots, u_P; \mathscr{E})$. Typical results obtained by averaging properties over these trajectories are pre-

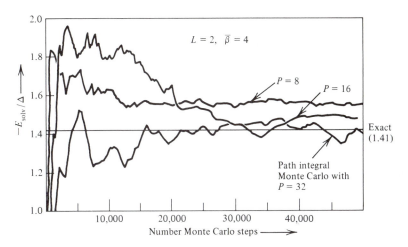

Fig. 6.6. Cumulative average for the solvation energy as computed by path integral quantum Monte Carlo.

sented in Fig. 6.6. The particular property considered in the figure is the solvation energy corresponding to the average of the coupling between the electric field and the quantal dipole. That is,

$$E_{solv} = \left\langle \frac{1}{P} \sum_{i=1}^{P} \mu \mathscr{E} u_i \right\rangle.$$

Figure 6.6 illustrates the convergence of the averaging as it depends upon the finite number of simulation steps and upon the finite number of discretized path integral points. In both instances, exact results are obtained in principle only when these numbers tend to infinity.

Note that the model is completely characterized by two dimensionless quantities: the reduced reciprocal temperature

$$\bar{\beta} = \beta \Delta,$$

and the localization parameter

$$L = \sigma \mu^2 / \Delta.$$

As discussed in Chapter 5, the model exhibits a localization transition-like behavior when $L > 1$. In Exercise 6.14, you are asked to run the quantum Monte Carlo code and attempt to observe this phenomenon.

Exercise 6.9 Verify that the exact result for the average coupling energy in this model is given by

$$E_{solv} = \frac{-\bar{\beta}\Delta \int_{-\infty}^{\infty} d\tilde{\mathscr{E}}\, e^{-\bar{\beta}\tilde{\mathscr{E}}^2/2L}\tilde{\mathscr{E}}^2 \sinh\left(\bar{\beta}\sqrt{1+\tilde{\mathscr{E}}^2}\right)}{\int_{-\infty}^{\infty} d\tilde{\mathscr{E}}\, e^{-\bar{\beta}\tilde{\mathscr{E}}^2/2L}\tilde{\mathscr{E}}^2 \cosh\left(\bar{\beta}\sqrt{1+\tilde{\mathscr{E}}^2}\right)}.$$

The computer code that generated these results for $P = 32$ points on the quantum path appears at the end of this chapter.

Additional Exercises

For those problems requiring the accumulation of statistics, you may want to use a compiled BASIC code, or rewrite the programs to run on a computer faster than the IBM PC. Also note that the largest integer carried by the IBM PC is 32767. Thus, when running ambitious projects on that machine you will want to alter the step counter to list hundreds of moves rather than singles.

6.10. Construct a Metropolis Monte Carlo code for a two-dimensional Ising magnet with $20 \times 20 = 400$ spins and periodic boundary conditions. Run the code and collect statistics from the trajectory to analyze the spin statistics. In particular, compute

$$\langle s_1 \rangle = \lim_{T \to \infty} \left\langle \frac{1}{N} \sum_{i=1}^{N} s_i \right\rangle_T$$

and for various separations between spins i and j, compute the correlation function

$$\langle s_i s_j \rangle - \langle s_i \rangle \langle s_j \rangle = \lim_{T \to \infty} \left[\left\langle \frac{1}{N_{ij}} \sum_{lm}^{(i,j)} s_l s_m \right\rangle_T - \langle s_i \rangle_T^2 \right],$$

where $\sum_{lm}^{(ij)}$ is over all pairs of spins in the lattice for which l and m are separated by the same distance as i and j, and N_{ij} is the total number of such pairs in the lattice. Perform these calculations for temperatures both above and below the critical point. Try to demonstrate the existence of long range correlations near the critical point and spontaneous symmetry breaking below the critical temperature.

6.11. Consider the two-dimensional Ising model with an additional external field in the energy,

$$\sum_{i=1}^{N} h_i s_i,$$

where $h_i = +h$ for spins lying on the left half of the top row of the square lattice, $h_i = -h$ for spins in the right half of the top row, and $h_i = 0$ for all other sites. For large enough h, and for $T < T_c$, this field will bias interfaces to exist near the middle and side columns of the square lattice.

(a) Modify the Monte Carlo code given in the Appendix to include this external field and use it to observe fluctuations of the interfaces.

(b) With the modified code at a temperature well below T_c (e.g., $T \approx \frac{1}{2} T_c$), determine the spin-spin correlation function for pairs of spins situated in the column midway between the left and right columns of the square lattice (i.e., the tenth column in the 20×20 lattice).

(c) Perform the same Monte Carlo calculation as in part (b) but now for the fifth column from the left.

(d) Plot the $\langle s_i s_j \rangle - \langle s_i \rangle \langle s_j \rangle$ determined in parts (b) and (c) as a function of the distance between the spins. Comment on your observations. How would your results change if the system size was changed to a 40×40 lattice, and the columns being sampled were changed to the tenth and twentieth, respectively?

6.12. Your computer contains a pseudo-random-number generator that creates (nearly) random sequences of numbers, x, distributed uniformly between 0 and 1. Develop an algorithm that uses this random number generator to create a Gaussian distribution of random numbers. A Gaussian distribution is

$$p(x) = \sqrt{\alpha/\pi}\, e^{-\alpha x^2},$$

and its first several moments are

$$\langle x \rangle = \langle x^3 \rangle = 0,$$

$$\langle x^2 \rangle = \frac{1}{2\alpha},$$

$$\langle x^4 \rangle = 3 \langle x^2 \rangle^2.$$

Study the numerical accuracy and convergence with which your algorithm reproduces these moments. (Note: There is more than one way to work out this exercise. One method uses an acceptance-rejection procedure like the Metropolis Monte Carlo algorithm. Another perhaps more efficient procedure employs a change in variables.)

6.13. In a Monte Carlo trajectory there are, in general, two steps to every successful move. The first stage is to make a trial move and the second stage tests to see if that move should be accepted. Therefore, the transition probability $w_{vv'}$ can be written

$$w_{vv'} = \pi_{vv'} \times A_{vv'}$$

where

$\pi_{vv'}$ = probability in a given step that if the system is in state v, it will make a *trial* transition to state v',

and

$A_{vv'}$ = probability that, if the system has made a trial move from v to v', the move will be accepted.

(a) Given an arbitrary form for $\pi_{vv'}$, write the form for $A_{vv'}$ that maintains detailed balance and such that either one or the other of $A_{vv'}$ and $A_{v'v}$ is equal to unity but not both.

(b) Consider the system with energy levels

$$E = \hbar\omega(v + \tfrac{1}{2}), \qquad v \text{ an integer}$$

and transition probabilities

$$\pi_{vv'} = \begin{cases} p, & v' = v + 1, \\ 1 - p, & v' = v - 1, \end{cases}$$

and for all other v and v', $\pi_{vv'}$ is zero. Find the value for p that minimizes the probability that if a system is in state v, it will make *no* transition. A good choice of p will make this probability zero, in which case every move will be accepted. Schemes such as this are often used in Monte Carlo calculations since they keep the system moving, thus reducing the length of a calculation required to obtain good statistics. This particular method is a simplified version of the scheme referred to as a "force bias" Monte Carlo or "smart" Monte Carlo.

6.14. Consider the two-state quantum system coupled to the Gaussian electric field $\mathscr{E}$ discussed in Secs. 5.8 and 6.4, and imagine applying a static non-fluctuating field $\mathscr{E}_{app}$. The total Hamiltonian is then

$$\mathscr{H} = \mathscr{H}_0 - m(\mathscr{E} + \mathscr{E}_{app}) + \mathscr{E}^2/2\sigma$$

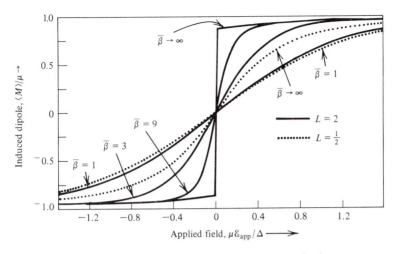

Fig. 6.7. Localization of a quantum fluctuating dipole.

where

$$\mathcal{H}_0 = \begin{bmatrix} 0 & -\Delta \\ -\Delta & 0 \end{bmatrix}$$

and

$$m = \begin{bmatrix} \mu & 0 \\ 0 & -\mu \end{bmatrix}.$$

The quantities Δ, μ, and are σ are constants.

(a) For this system, show that the average dipole, $\langle m \rangle$, is given by

$$\langle m \rangle = \frac{\displaystyle\int_{-\infty}^{\infty} d\bar{\mathscr{E}} \exp\left(-\bar{\beta}\bar{\mathscr{E}}^2/2L\right) \sinh\left(\bar{\beta}\xi\right)(\bar{\mathscr{E}} + \bar{\mathscr{E}}_{\text{app}})/\xi}{\displaystyle\int_{-\infty}^{\infty} d\bar{\mathscr{E}} \exp\left(-\bar{\beta}\bar{\mathscr{E}}^2/2L\right) \cosh\left(\bar{\beta}\xi\right)}$$

where $\xi^2 = [1 + (\bar{\mathscr{E}} + \bar{\mathscr{E}}_{\text{app}})^2]$, $\bar{\mathscr{E}}_{\text{app}} = (\mu/\Delta)\mathscr{E}_{\text{app}}$, $\bar{\beta} = \beta\Delta$, and $L = \sigma\mu^2/\Delta$.

(b) Numerically evaluate the integrals in part (a) and demonstrate the behavior of $\langle m \rangle$ illustrated in Fig. 6.7. (Note that as $\bar{\beta} \to \infty$, the integrals can be evaluated analytically by the method of steepest descent.) Note that for $L > 1$, the system exhibits a localization transition when $\bar{\beta} \to \infty$. This phenomenon was discussed in Sec. 5.8 and Exercise 5.26.

(c) Modify the Monte Carlo code given in Sec. 6.4 to study the localization phenomena. In particular, attempt to reproduce the results given in the figure by averaging over Monte Carlo trajectories.

6.15. Devise and implement a method for computing by Monte Carlo the *free* energy of solvation for the two-state quantum system coupled to the Gaussian electric field. [*Hint*: You will need to compute the solvation energy, E_{solv}, at a variety of values of the dipole magnitude μ keeping all other parameters fixed.] Compare the Monte Carlo results with the exact result obtained by performing the one-dimensional numerical integral

$$Q = (2\Delta/\mu) \int_{-\infty}^{\infty} d\bar{\mathscr{E}} \exp(-\bar{\beta} \, \bar{\mathscr{E}}^2/2L) \cosh(\bar{\beta}\sqrt{1 + \bar{\mathscr{E}}^2}).$$

6.16. Modify the Ising model Monte Carlo code for 400 spins to perform the umbrella sampling calculation of $\bar{A}(M)$ as described in Sec. 6.3. Perform these calculations at a variety of temperatures and use the value of M at which $\bar{A}(M)$ is a minimum as an estimate of the spontaneous magnetization. Compare your results so obtained with the exact result for the spontaneous magnetization of the infinite square two-dimensional Ising model. The exact result (the derivation of which was first given by C. N. Yang in the 1950s) is

$$m(T) = \begin{cases} 0, & T > T_c, \\ (1 + z^2)^{1/4}(1 - 6z^2 + z^4)^{1/8}(1 - z^2)^{-1/2}, & T < T_c, \end{cases}$$

where $z = \exp(-2\beta J)$, and T_c corresponds to $z_c = \sqrt{2} - 1$.

Bibliography

The literature contains several helpful reviews of Monte Carlo methods. For example:

K. Binder in *Monte Carlo Methods in Statistical Physics,* ed. by K. Binder (Springer-Verlag, N.Y., 1979).

K. Binder in *Applications of the Monte Carlo Method,* ed. by K. Binder (Springer-Verlag, N.Y., 1983).

J. P. Valleau and S. G. Whittington, in *Statistical Mechanics Part A,* ed. by B. J. Berne (Plenum, N.Y., 1977).

J. P. Valleau and G. M. Torrie, *ibid.*

Further discussion of Monte Carlo methods together with software designed for microcomputers is found in

S. E. Koonin, *Computational Physics* (Benjamin-Cummings, Menlo Park, Calif., 1986).

Koonin also provides an appendix with a synopsis of the BASIC language.

While we have not discussed molecular dynamics, two reviews of that technique are

A. Rahman in *Correlation Functions and Quasi-Particle Interactions in Condensed Matter,* ed. by J. W. Halley (Plenum, N.Y., 1978).

I. R. McDonald in *Microscopic Structure and Dynamics of Liquids,* ed. by J. DuPuy and A. J. Dianoux (Plenum, N.Y., 1977).

Working with and observing the trajectories of the Monte Carlo code described in this chapter may help develop an appreciation and intuition for interfacial fluctuations in a two-phase system. These fluctuations are sometimes called *capillary waves.* Treatments of surface fluctuations employing lattice gases and Monte Carlo simulation of Ising models are found in

S. K. Ma, *Statistical Mechanics* (World Scientific, Philadelphia, 1985).

Appendix

Monte Carlo Program for the Two-Dimensional Ising Model*

```
10    DEFINT A,S,I,J,K,M,N
20    DIM A(22,22),SUMC(5),KI(5),KJ(5)
30 ON KEY(1) GOSUB 40
40    ICOUNT=0 'initialize counter
50 CLS:KEY OFF
60 LOCATE 25,50: PRINT"-PRESS F1 TO RESTART-"
70 COLOR 15,0: LOCATE 2,15: PRINT "MONTE CARLO ISING MODEL "
80 COLOR 7: PRINT:PRINT "Monte Carlo Statistics for a 20X20 ISING MODEL with"
90 PRINT "            periodic boundary conditions."
100 PRINT: PRINT" The critical temperature is approximately 2.0."
110 PRINT:PRINT "CHOOSE THE TEMPERATURE FOR YOUR RUN.  Type a number between "
120 INPUT"     0.1 and 100, and then press  'ENTER'.",T
130 IF T<.1 THEN T=.1 ELSE IF T>100 THEN T=100
140 PRINT ">>>>> temperature=" T: T=1/T
150 KEY(1) ON
160 PRINT:PRINT "DO YOU WANT TO STUDY THE CORRELATION FUNCTION (Y OR N)?"
170 COR$=INPUT$(1)
180 IF COR$="y" THEN COR$="Y"
190 IF COR$="Y" THEN PRINT ">>>> correlation data will be shown" ELSE PRINT
             >>>> no correlation data will be shown"
200 PRINT:PRINT "PICK THE TYPE OF INITIAL SPIN CONFIGURATION"
210 PRINT,"TYPE c FOR 'CHECKERBOARD' PATTERN, OR"
220 PRINT,"TYPE i FOR 'INTERFACE' PATTERN"
230 PRINT,"TYPE u FOR 'UNEQUAL INTERFACE' PATTERN"
240 X$=INPUT$(1)
250 IF X$="C" OR X$="c" GOTO 370
260 IF X$="u" THEN X$="U"
270 IF X$="i" THEN X$="I"
280 IF X$="I" OR X$="U" THEN 290 ELSE 210 'ROUTING TO PROPER INITIAL SETUP
290 CLS 'initial INTERFACE setup
300    IF X$="U" THEN MAXJ=14 ELSE MAXJ=10
310    FOR I=0 TO 22
320    FOR J=0 TO MAXJ: A(I,J)=+1: NEXT
330    FOR J=MAXJ+1 TO 22: A(I,J)=-1: NEXT
340    A(I,0)=-1: A(I,21)=1:
350    NEXT
360 GOTO 420
370 CLS 'INITIAL checkerboard PATTERN
380    A(0,0)=1
390    FOR I=0 TO 20: A(I+1,0)=-A(I,0)
400    FOR J=0 TO 20: A(I,J+1)=-A(I,J):NEXT
410    NEXT
420 REM initial display:
430        LOCATE 25,50:PRINT"-PRESS F1 TO RESTART-"
440        FOR I=1 TO 20
450        FOR J=1 TO 20
```

* This program was written by Elaine Chandler.

```
460        FOR JZ=2*J-1 TO 2*J
470        LOCATE I,JZ: IF A(I,J)=1 THEN PRINT CHR$(219) ELSE PRINT CHR$(176)
480        NEXT JZ,J,I
490        LOCATE 10,50: PRINT"TEMP="1/T
500 TIME$="00:00:00"
510 IF X$="U" THEN NPLUS=280 ELSE NPLUS=200
520 IF COR$="Y" THEN GOSUB 710
530 M=INT(20*RND+1):N=INT(20*RND+1): S=-A(N,M): ICOUNT=ICOUNT+1 '**flip a spin
540 B=T*S*(A(N-1,M)+A(N+1,M)+A(N,M-1)+A(N,M+1))*2
550 IF EXP(B)<RND GOTO 620                         'test against random#
560 A(N,M)=S: NPLUS=NPLUS+S
570 IF N=1 THEN A(21,M)=S ELSE IF N=20 THEN A(0,M)=S
580 IF M=1 THEN A(N,21)=S ELSE IF M=20 THEN A(N,0)=S
590    FOR IX=2*M-1 TO 2*M  'update the display
600    LOCATE N,IX:IF A(N,M)=1 THEN PRINT CHR$(219) ELSE PRINT CHR$(176)
610    NEXT
620 LOCATE 23,21: PRINT ICOUNT: LOCATE 23,30: PRINT TIME$
630 IF (ICOUNT MOD 100)=0 THEN GOSUB 670
640 IF COR$="Y" AND (ICOUNT MOD 400)=0 THEN GOSUB 750
650 GOTO 530
660 END
670 LOCATE 12,50: PRINT "AT "ICOUNT;
680 XN=NPLUS/400!
690 PRINT USING "  N+/N=.###";XN
700 RETURN
710 LOCATE 14,47: PRINT "Correlation Function:"
720 LOCATE 15,51: PRINT "d  <s(0)s(d)>": LOCATE 16,50: PRINT"-------------"
730 GOSUB 750
740 RETURN
750 FOR M=1 TO 5: SUMC(M)=0:    'Correlation calculation
760 LOCATE 14,69: PRINT"(at "ICOUNT")"
770 FOR I=1 TO 20
780 FOR J=1 TO 20:KJ=(J+M) MOD 20: KI=(I+M) MOD 20: CC%=A(KI,J)+A(I,KJ)
790 IF CC%=0 THEN GOTO 810
800 SUMC(M)=SUMC(M)+A(I,J)*CC%
810 NEXT J,I
820 LOCATE 16+M,50: PRINT M: LOCATE 16+M,54: PRINT USING "+#.###"; SUMC(M)/800
830 NEXT M: RETURN
```

Quantum Monte Carlo Program for a Two-Level System Coupled to a Fluctuating Field*

```
10 REM PROGRAM TO SIMULATE A TWO-LEVEL SYSTEM  COUPLED TO
20 REM AN ADIABATIC GAUSSIAN FLUCTUATING FIELD
30 WHITE$ = CHR$(219)+CHR$(219)
40 BLACK$ = CHR$(176)+CHR$(176)
50 ON KEY(2) GOSUB 90
60 KEY(2) ON
70 DIM SIGMA%(32)
80 DIM ESOL!(2000)
90 REM RESTART FROM HERE
100 CLS
110 RANDOMIZE(310941!)
120 KEY(2) ON
130 LOCATE 1,1
140 PRINT "ENTER THE LOCALIZATION PARAMETER 'L'"
150 INPUT "***BETWEEN 0.01 AND 10 ***";LOCAL!
160 IF LOCAL!<.01 GOTO 150
170 IF LOCAL!>10 GOTO 150
180 PRINT "ENTER THE REDUCED TEMPERATURE 'BETA'";
190 INPUT "***BETWEEN 0.01 AND 16 ***";RBETA!
200 IF RBETA!<.01 GOTO 190
210 IF RBETA!>16 GOTO 190
220 INPUT "TOTAL NUMBER OF STEPS (INTEGER BELOW 1E+20)";MOVE!
230 IF MOVE!<0 THEN GOTO 220
240 IF MOVE!<>INT(MOVE!) THEN GOTO 220
250 IF MOVE!>1E+20 THEN GOTO 220
260 IVAR!=RBETA!/(2*LOCAL!)
270 FSTEP!=SQR(1/IVAR!)
280 IF FSTEP!>1 THEN FSTEP!=1
290 KAY!=-.5*LOG(RBETA!/32)
300 TKAY!=KAY!*2
310 LAMBDA!=RBETA!/32
320 TLAMBDA!=2*LAMBDA!
330 NEWENERGY!=0
340 FILD!=0
350 ESOLV!=0
360 NEWESOLV!=0
370 FOR I=1 TO 32
```

* This program was written by Faramarz Rabii.

```
390 NEXT I
400 CLS
410 GOSUB 1440
420 GOSUB 980
430 GOSUB 860
440 FOR IRUN!=1 TO MOVE!
450 LOCATE 16,45
460 PRINT "STEP =";IRUN!
470 JAY%=INT(RND*39)+1
480 IF JAY%>32 THEN GOTO 520
490 OLDSIGMA%=SIGMA%(JAY%)
500 IF OLDSIGMA%=1 THEN SIGMA%(JAY%)=-1 ELSE SIGMA%(JAY%)=1
510 GOTO 550
520 GOSUB 1320
530 OLDFILD!=FILD!
540 FILD!=FILD!+FSTEP!*(.5-RND)
550 OLDENERGY!=NEWENERGY!
560 OLDESOLV!=NEWESOLV!
570 GOSUB 1050
580 IF NEWENERGY!>=OLDENERGY! THEN GOTO 610
590 PROB!=EXP(NEWENERGY!-OLDENERGY!)
600 IF RND>PROB! THEN GOTO 630
610 IF JAY%<33 THEN DIP!=NEWDIP!
620 GOTO 660
630 IF JAY%>32 THEN FILD!=OLDFILD! ELSE SIGMA%(JAY%)=OLDSIGMA%
640 NEWENERGY!=OLDENERGY!
650 NEWESOLV!=OLDESOLV!
660 GOSUB 1240
670 ESOLV!=ESOLV!+NEWESOLV!
680 TST!=IRUN!/100
690 IF TST!<>INT(TST!) THEN GOTO 830
700 ESOL!(TST!)=ESOLV!/IRUN!
710 LOCATE 17,22
720 PRINT "                "
730 LOCATE 17,3
740 PRINT "SOLVATION ENERGY = ";-NEWESOLV!/RBETA!
750 LOCATE 18,3
760 PRINT "(IN UNITS OF DELTA)"
770 LOCATE 19,14
780 PRINT "           "
790 LOCATE 19,3
800 PRINT"DIPOLE = ";DIP!/32
810 LOCATE 20,3
820 PRINT "(IN UNITS OF MU)"
830 IF IRUN!>MOVE! THEN GOTO 1770
840 NEXT IRUN!
850 GOTO 1770
860 REM OUT PUT CONFIGURATION
870 DIP!=0
880 LOCATE 11,3
890 FOR INDEX%=1 TO 32
900 IF SIGMA%(INDEX%)=1 THEN PRINT BLACK$; ELSE PRINT WHITE$;
910 DIP!=DIP!+SIGMA%(INDEX%)
920 NEXT INDEX%
930 LOCATE 12,3
940 FOR INDEX%=1 TO 32
950 IF SIGMA%(INDEX%)=1 THEN PRINT WHITE$; ELSE PRINT BLACK$;
960 NEXT INDEX%
970 RETURN
980 REM subsection to compute overall initial energy
990 NEWENERGY!=0
1000 FOR I=1 TO 32
1010 IF I=32 THEN J=1 ELSE J=I+1
1020 NEWENERGY!=NEWENERGY!+KAY!*SIGMA%(I)*SIGMA%(J)
1030 NEXT I
1040 RETURN
1050 REM SUBSECTION TO QUICKLY COMPUTE NEW ENERGY
1060 IF JAY%=1 THEN JAYM1%=32 ELSE JAYM1%=JAY%-1
1070 IF JAY%=32 THEN JAYP1%=1 ELSE JAYP1%=JAY%+1
1080 IF JAY%>32 THEN GOTO 1200
1090 IF SIGMA%(JAY%)=1 THEN GOTO 1150
1100 NEWESOLV!=OLDESOLV!-FILD!*TLAMBDA!
1110 NEWENERGY!=OLDENERGY!+NEWESOLV!
1120 NEWENERGY!=NEWENERGY!-TKAY!*(SIGMA%(JAYP1%)+SIGMA%(JAYM1%))
1130 NEWDIP!=DIP!-2
1140 GOTO 1230
1150 NEWESOLV!=OLDESOLV!+FILD!*TLAMBDA!
1160 NEWENERGY!=OLDENERGY!+NEWESOLV!
1170 NEWENERGY!=NEWENERGY!+TKAY!*(SIGMA%(JAYP1%)+SIGMA%(JAYM1%))
1180 NEWDIP!=DIP!+2
1190 GOTO 1230
1200 NEWENERGY!=OLDENERGY!+IVAR!*(OLDFILD!*OLDFILD!-FILD!*FILD!)
1210 NEWESOLV!=LAMBDA!*DIP!*FILD!
1220 NEWENERGY!=NEWENERGY!-OLDESOLV!+NEWESOLV!
1230 RETURN
1240 REM update display
1250 IF (IRUN!/4)<>INT((IRUN!/4)) THEN RETURN
1260 IF JAY%>32 THEN GOTO 1310
1270 LOCATE 11,2*JAY%+1
1280 IF SIGMA%(JAY%)=1 THEN PRINT BLACK$ ELSE PRINT WHITE$
```

```
1290 LOCATE 12,2*JAY%+1
1300 IF SIGMA%(JAY%)=1 THEN PRINT WHITE$ ELSE PRINT BLACK$
1310 RETURN
1320 REM SUBSECTION TO OUT-PUT EXTERNAL FIELD
1330 OLDFLDMAG%=FLDMAG%
1340 FLDMAG%=-INT(FILD!)
1350 IF ABS(FLDMAG%)>9 THEN FLDMAG%=9*SGN(FLDMAG%)
1360 IF ABS(FLDMAG%)>=ABS(OLDFLDMAG%) THEN GOTO 1400
1370 LOCATE OLDFLDMAG%+11,74
1380 PRINT CHR$(219);CHR$(219);CHR$(219)
1390 GOTO 1430
1400 LOCATE FLDMAG%+11,74
1410 PRINT CHR$(176);CHR$(176);CHR$(176)
1420 GOTO 1430
1430 RETURN
1440 REM SUBSECTION TO DRAW A BOX AROUND THE FIELD OUT-PUT
1450 FOR I=1 TO 19
1460 LOCATE I+1,73
1470 PRINT CHR$(179);CHR$(219);CHR$(219);CHR$(219);CHR$(179)
1480 NEXT I
1490 LOCATE 1,73
1500 PRINT "-----"
1510 LOCATE 21,73
1520 PRINT "-----"
1530 LOCATE 11,70
1540 PRINT "0.0"
1550 LOCATE 2,70
1560 PRINT "9.0"
1570 LOCATE 20,69
1580 PRINT "-9.0"
1590 LOCATE 11,74
1600 PRINT CHR$(176);CHR$(176);CHR$(176)
1610 LOCATE 22,75
1620 PRINT CHR$(24)
1630 LOCATE 23,63
1640 PRINT "FLUCTUATING FIELD";
1650 LOCATE 23,9
1660 PRINT "TO RESTART PRESS F2"
1670 REM DISPLAY INITIAL CONDITIONS
1680 LOCATE 3,1
1690 PRINT "LOCALIZATION PARAMETER L=";LOCAL!
1700 PRINT
1710 PRINT "REDUCED TEMPERATURE BETA=";RBETA!
1720 LOCATE 9,29
1730 PRINT "THE QUANTUM PATH"
1740 LOCATE 10,36
1750 PRINT CHR$(25)
1760 RETURN
1770 REM OUT-PUT SOLVATION ENERGY VALUES
1780 LPRINT "RESULTS FOR THE SIMULATION OF A TWO-LEVEL SYSTEM COUPLED"
1790 LPRINT "TO AN ADIABATIC FIELD."
1800 LPRINT
1810 LPRINT "INITIAL CONDITIONS ARE:"
1820 LPRINT
1830 LPRINT "LOCALIZATION PARAMETER  L = ";LOCAL!
1840 PRINT
1850 LPRINT "REDUCED TEMPERATURE BETA = ";RBETA!
1860 LPRINT
1870 LPRINT "#OF STEPS","AVERAGE SOLVATION ENERGY IN UNITS OF DELTA"
1880 LPRINT
1890 IMAX%=MOVE!/100
1900 FOR I=1 TO IMAX%
1910 LPRINT I*100,-ESOL!(I)/RBETA!
1920 NEXT I
1930 LOCATE 23,1
1940 END
```

CHAPTER 7

Classical Fluids

The relative arrangements of atoms and molecules in fluid and solid phases are most often accurately described in terms of the principles of classical statistical mechanics. There are, of course, electrons surrounding the nuclei in these systems, and the behavior of electrons is surely quantum mechanical in nature. Yet after averaging over these quantal fluctuations, the remaining problem is to sample the statistical configurations of the nuclei in the effective interactions induced by the electrons we have already integrated out. An example of these effective interactions are the Born–Oppenheimer potentials considered in Chapter 4.

Schematically, this procedure is as follows. In the partition function

$$Q = \sum_v \exp\left(-\beta E_v\right),$$

the state v can be characterized by the configuration of the nuclei, denoted by the symbol R, and the electronic state i, parameterized by R. It is then convenient to factor or partition the states or fluctuations according to the configurations of the nuclei,

$$Q = \sum_R \left\{ \sum_{i(R)} \exp\left[-\beta E_{R,i(R)}\right] \right\}$$

$$\equiv \sum_R \exp\left[-\beta \tilde{E}_R\right],$$

where $i(R)$ stands for the ith state of the electrons when the configuration of the nuclei (the centers of the atoms) is held fixed at

R (this configurational variable, R, is actually an enormous collection of all the coordinates necessary to specify the locations of all the atomic centers). The quantity $\tilde{E}_R$ is obtained by performing the Boltzmann weighted sum within the curly brackets. It is the effective energy or free energy governing the statistics for the configurations of the nuclei.

The discussion in the previous paragraph is, of course, highly schematic. We do know, however, from Sec. 5.8 and 6.4, that at least one way to perform the Boltzmann weighted sum over the electronic fluctuations or states is to evaluate quantum path summations. The sum over $i(R)$ then denotes such a procedure. In general, we see that $\tilde{E}_R$ is a free energy that depends upon temperature. Often, however, the electronic states of a system are dominated by the lowest energy level. In that case, the result of averaging out the quantal fluctuations of electrons yields an $\tilde{E}_R$ that must be the ground state Born–Oppenheimer energy surface for all the nuclei. To simplify our considerations, we will assume that this ground state dominance is an accurate approximation for the systems we examine in this chapter.

The remaining problem is that of studying the spatial configurations of the nuclei. This problem is usually well approximated with a classical mechanical model. The reason, made more precise later in this chapter, is that the nuclei are much heavier than electrons. The relatively high mass implies that the quantum uncertainties in the positions of nuclei are relatively small, and as a result, quantum dispersion (i.e., the width of wavefunctions) becomes unimportant when considering the spatial fluctuations of the nuclei in these systems.

An important exception where a classical model of a fluid is not acceptable is low temperature helium. We will not discuss this type of system. Instead, we will consider the meaning of velocity distributions and intermolecular structure for fluids like argon or benzene or water. Here, after averaging out the fluctuations associated with electronic states, classical models are accurate.

7.1 Averages in Phase Space

When adopting a classical model, the microscopic state of the system is characterized by a point in *phase space*. That is, a state is specified by listing the coordinates and conjugate momenta of all the classical degrees of freedom in the system:

$$(\mathbf{r}_1, \mathbf{r}_2, \ldots, \mathbf{r}_N; \mathbf{p}_1, \mathbf{p}_2, \ldots, \mathbf{p}_N) = (r^N, p^N)$$
$$= \text{point in phase space for an } N \text{ particle system.}$$

Here $\mathbf{r}_i$ = position of particle i, and $\mathbf{p}_i$ = momentum of particle i, and r^N and p^N are abbreviations for points in configuration space and momentum space, respectively.

In order to perform statistical mechanical calculations on a classical model, we must be able to compute the classical analog of objects like the canonical partition function,

$$Q(\beta, N, V) = \sum_v e^{-\beta E_v}.$$

The energy associated with a point in phase space is the Hamiltonian. That is,

$$E_v \rightarrow \mathscr{H}(r^N, p^N) = K(p^N) + U(r^N),$$

where $K(p^N)$ denotes the kinetic energy of the classical degrees of freedom, and $U(r^N)$ is the potential energy. This latter part to the energy is obtained by averaging over all the quantum degrees of freedom that are not treated explicitly in the classical model. In other words, the potential energy function $U(r^N)$ must be determined from a quantum electronic structure calculation. Finally, we remark that in a conservative Newtonian system, the kinetic energy is a function of momenta only, and the potential energy is a function of coordinates only.*

Since the points in phase space form a continuum, the classical canonical partition function must be something like

$$Q = (?) \int dr^N \int dp^N \exp\left[-\beta\mathscr{H}(r^N, p^N)\right],$$

where $\int dr^N \int dp^N$ is an abbreviation for

$$\int d\mathbf{r}_1 \int d\mathbf{r}_2 \cdots \int d\mathbf{r}_N \int d\mathbf{p}_1 \int d\mathbf{p}_2 \cdots \int d\mathbf{p}_N.$$

But the phase space integral has dimensions of action to the DN power (D = dimensionality). Thus, there must be a multiplicative factor, denoted by (?) in the equation, which makes Q dimensionless. It should be a universal factor. As a result, we can determine it by studying in detail one particular system. For an ideal gas of structureless particles,

$$\mathscr{H}(r^N, p^N) = \sum_{i=1}^{N} p_i^2/2m.$$

---

* Classical Lagrangian systems with holonomic constraints do have kinetic energies that can depend upon configurations. However, such a dependence is an artifact of the holonomic constraints that are absent for any system found in nature.

Hence,

$$Q = (?)V^N \left[\int d\mathbf{p} \exp(-\beta p^2/2m) \right]^N.$$

By comparing this result with what we calculated in Sec. 4.7 (our discussion of the classical ideal gas), we find

$$(?) = (N!\, h^{3N})^{-1}.$$

Thus,

$$Q = \sum_v e^{-\beta E_v} \rightarrow (1/N!\, h^{3N}) \int dr^N \int dp^N \exp[-\beta \mathcal{H}(r^N, p^N)].$$

We can understand the factor of $N!$ by noting that the N identical particles should be indistinguishable. Hence, the phase space integral overcounts states $N!$ times (the number of different ways we can relabel all the particles). To avoid the overcounting, we must divide by $N!$. The factor of h^{3N} occurs for a less transparent reason (except that it has dimensions of action to the $3N$ power). A rough argument focuses attention on the uncertainty principle, $\delta r^N \delta p^N \sim h^{3N}$. Hence, we expect our differential volume elements in phase space to scale like h^{3N}. That is,

$$\sum_v = (N!)^{-1} \sum_{\delta r^N, \delta p^N} \rightarrow (N!)^{-1} h^{-3N} \int dr^N \, dp^N.$$

Exercise 7.1 For a system composed of three distinguishable types of particles, A, B, and C, justify the classical partition function formula

$$Q = [(N_A!)(N_B!)(N_C!) h^{3(N_A+N_B+N_C)}]^{-1} \int dr^N \int dp^N$$
$$\times \exp[-\beta \mathcal{H}(r^N, p^N)],$$

where (r^N, p^N) is an abbreviation for a point in phase space for the $(N_A + N_B + N_C)$-particle system.

Often, we have systems in which there are quantum degrees of freedom in each atom that do not couple to the classical variables and therefore do not affect $U(r^N)$. In that case

$$Q = \frac{1}{N!\, h^{3N}} q_{\text{quantum}}^N(\beta) \int dr^N \int dp^N \exp[-\beta \mathcal{H}_{\text{classical}}],$$

where $q_{\text{quantum}}^N(\beta)$ is the partition function for those uncoupled quantum mechanical degrees of freedom.

The probability of a state in a classical system is $f(r^N, p^N)\, dr^N\, dp^N$, where

$f(r^N, p^N)$ = probability distribution for observing a system at phase space point (r^N, p^N).

Clearly,

$$f(r^N, p^N) = \exp\left[-\beta \mathcal{H}(r^N, p^N)\right] \Big/ \int dr^N \int dp^N \exp\left[-\beta \mathcal{H}(r^N, p^N)\right].$$

Since the Hamiltonian breaks into two parts, $K(p^N)$ and $U(r^N)$, the phase space distribution factors as

$$f(r^N, p^N) = \Phi(p^N)P(r^N),$$

where

$$\Phi(p^N) = \exp\left[-\beta K(p^N)\right] \Big/ \int dp^N \exp\left[-\beta K(p^N)\right]$$

= probability distribution for observing system at momentum space point p^N

and

$$P(r^N) = \exp\left[-\beta U(r^N)\right] \Big/ \int dr^N \exp\left[-\beta U(r^N)\right]$$

= probability distribution for observing system at configuration space point r^N.

Exercise 7.2 Derive this result.

Exercise 7.3 Show that canonical partition function factors in such a way that

$$Q = Q_{\text{ideal}} Q_{\text{con}},$$

where Q_{ideal} is the ideal gas partition function and

$$Q_{\text{con}} = V^{-N} \int dr^N \exp\left[-\beta U(r^N)\right].$$

Further factorization of the momentum distribution is possible since the kinetic energy is a sum of single particle energies $\sum_i p_i^2/2m$. Thus

$$\Phi(p^N) = \prod_{i=1}^{N} \phi(\mathbf{p}_i),$$

where

$$\phi(\mathbf{p}_i) = e^{-(\beta p_i^2/2m)} \Big/ \int d\mathbf{p}\, e^{-\beta p^2/2m}.$$

[Note, $p_i^2 = (p_{ix}^2 + p_{iy}^2 + p_{iz}^2)$, where $p_{i\alpha}$ is the Cartesian component of the momentum for particle i in direction α.] The single particle momentum distribution, $\phi(\mathbf{p})$, is usually called the Maxwell-Boltzmann (MB) distribution. It is the correct momentum distribution function for a particle of mass m in a thermally equilibrated system. The system can be in any phase (gas, liquid, or solid) and the distribution is still valid provided classical mechanics is accurate. One consequence is that the average speed (or momentum) of a particle is the same in a liquid and a gas (provided the temperature is the same). Of course, the frequency of collisions in a liquid is much higher than that in gas. For this reason, a molecule will travel much farther per unit time in a gas phase than in a condensed phase even though the single molecule velocity distributions are identical in the two phases.

Some typical calculations we could perform with the MB distribution are

$$\langle |\mathbf{p}| \rangle = \frac{\int d\mathbf{p}\, |\mathbf{p}| \exp(-\beta p^2/2m)}{\int d\mathbf{p} \exp(-\beta p^2/2m)} = \frac{4\pi \int_0^\infty dp\, p^3 \exp(-\beta p^2/2m)}{\left[\int_{-\infty}^\infty dp \exp(-\beta p^2/2m)\right]^3}$$

$$= (8k_B Tm/\pi)^{1/2}$$

and

$$\langle p^2 \rangle = \langle p_x^2 \rangle + \langle p_y^2 \rangle + \langle p_z^2 \rangle = 3\langle p_x^2 \rangle = 3mk_B T.$$

Exercise 7.4 Derive these results.

Exercise 7.5 Show that the classical partition function is

$$Q = (N!)^{-1}\lambda_T^{-3N} \int dr^N \exp[-\beta U(r^N)],$$

where

$$\lambda_T = h/\sqrt{2\pi mk_B T}$$

is called the thermal wavelength.

Exercise 7.6 Given that at typical liquid densities $\rho\sigma^3 \approx 1$ ($\rho = N/V$, and σ is the diameter of a molecule), estimate the mean free path of and collision frequency of a molecule in a room temperature liquid. Compare these numbers with gas phase values.

We can use the MB distribution to assess whether it is a good approximation to avoid Schrödinger's equation and use classical mechanics to describe the microscopic states of a system. Classical mechanics is accurate when the De Broglie wavelength,

$$\lambda_{DB} = h/p,$$

is small compared to relevant intermolecular length scales. An estimate of a typical value for λ_{DB} is

$$\lambda_{DB} \sim h/\langle |p| \rangle = h/\sqrt{8k_B Tm/\pi} \approx \lambda_T.$$

This length is the typical distance over which the precise location of a particle remains uncertain due to Heisenberg's principle. When λ_T is small compared to any relevant length scale, the quantal nature of fluctuations becomes unimportant. For a dilute gas, relevant lengths are $\rho^{-1/3}$ (the typical distance between particles) and σ (the diameter of a particle). Here,

$$\lambda_T < \sigma$$

would seem to suffice as a criterion for classical statistical mechanics to be valid. In general, however, one must consider the distance scales that characterize the spatially smallest fluctuations under consideration. The classical model is valid where, for typical fluctuations, intermolecular distance changes of the order of λ_T produce potential energy changes that are small compared to $k_B T$. At liquid densities, therefore, the parameter that measures the importance of quantum mechanics is $\beta\lambda_T\langle |F| \rangle$, where $|F|$ is the magnitude of the force between neighboring pairs of particles. When this parameter is small, a classical model is accurate; when the parameter is large, the quantal nature of fluctuations must be accounted for.

Exercise 7.7 For liquid nitrogen at its triple point, compare λ_T with the "diameter" of the N_2 molecule (roughly 4 Å).

The pressure is obtained from the free energy by differentiating with respect to the volume V. That is, $p = -(\partial A/\partial V)_{T,N}$. Due to the factorization of the partition function, this relationship means that the equation of state is obtained from the configurational part of the partition function. That is

$$\beta p = (\partial \ln Q/\partial V)_{N,\beta}$$

$$= \frac{\partial}{\partial V} \ln \int dr^N \exp\left[-\beta U(r^N)\right].$$

We ignore all interactions of the boundary with the system other than those that confine the system to a particular volume in space. Hence, the volume dependence of the configurational integral is in the limits of integration only. Note also that as the volume increases, the configurational integral necessarily increases, too, since the integrand is always positive. Therefore, the derivative yielding p is always positive. Thus within the class of models we examine here, the pressure of an equilibrium system is always positive.

Notice that with classical models, we predict that the configurational part of the partition function is independent of the momenta and masses of particles. Therefore, the equation of state, $p = p(\beta, \rho)$, is independent of the masses of the particles in the system. Thus, if the translational and rotational motions of water molecules are well described by classical mechanics, then the equation of state of the liquids H_2O and D_2O will be the same. The experimental situation, however, is that there are noticeable differences. For example, the density maximum in $\rho(T)$ at $p = 1$ atm occurs at 4°C in H_2O, and at 10°C in D_2O. The freezing of D_2O occurs at a higher temperature than H_2O, too. To think about this phenomenon on physical grounds you can imagine that the role of quantum mechanics is to blur the positions of the atoms over a distance $\sqrt{\beta\hbar^2/m}$. For a proton at room temperature, this corresponds to a length of about 0.3 Å. The diameter of a water molecule is about 3 Å. Since the protons are strongly bound to oxygen atoms, most of the blurring is associated with librational motion of the water molecules. As the atomic mass increases, the diffusiveness of the location of the atoms diminishes, and the fluid becomes more ordered. This is the reason why, for example, D_2O ice melts at a higher temperature than H_2O ice.

7.2 Reduced Configurational Distribution Functions

The configurational distribution $P(r^N)$, does not factor into single particle functions because the potential energy, $U(r^N)$, couples together all of the coordinates. However, we can still discuss distribution functions for a small number of particles by integrating over all coordinates except those pertaining to the particles of interest. For example,

$$P^{(2/N)}(\mathbf{r}_1, \mathbf{r}_2) = \int d\mathbf{r}_3 \int d\mathbf{r}_4 \cdots \int d\mathbf{r}_N \, P(r^N)$$

$$= \text{joint probability distribution for finding}$$
$$\text{particle 1 at position } \mathbf{r}_1 \text{ and particle 2 at } \mathbf{r}_2.$$

This distribution function is called a *specific* probability distribution because it specifically requires particle 1 (and no other particle) to be at $\mathbf{r}_1$, and similarly, it must be particle 2 at $\mathbf{r}_2$. Such requirements are not physically relevant for systems composed of N indistinguishable particles. Further, the specific reduced distribution must be vanishingly small as N grows to a reasonable macroscopic value (say 10^{23}).[*]

The more meaningful quantities are *generic* reduced distribution functions. For example, let

$\rho^{(2/N)}(\mathbf{r}_1, \mathbf{r}_2) =$ joint distribution function for finding a particle (any one) at position $\mathbf{r}_1$, and any other particle (in the N particle system) at $\mathbf{r}_2$.

Note that there are N possible ways of picking the first particle (the one at $\mathbf{r}_1$), and there are $N-1$ ways of picking the second. Thus,

$$\rho^{(2/N)}(\mathbf{r}_1, \mathbf{r}_2) = N(N-1)P^{(2/N)}(\mathbf{r}_1, \mathbf{r}_2).$$

In general

$\rho^{(n/N)}(\mathbf{r}_1, \mathbf{r}_2, \ldots, \mathbf{r}_n) =$ joint distribution function that in an N particle system, a particle will be found at $\mathbf{r}_1$, another at $\mathbf{r}_2, \ldots$, and another at $\mathbf{r}_n$,

$$= [N!/(N-n)!] \int dr^{N-n}$$

$$\times \exp\left[-\beta U(r^N)\right] \Big/ \int dr^N \exp\left[-\beta U(r^N)\right],$$

where dr^{N-n} is an abbreviation for $d\mathbf{r}_{n+1}\, d\mathbf{r}_{n+2} \cdots d\mathbf{r}_N$. For an isotropic fluid, we have

$$\rho^{(1/N)}(\mathbf{r}_1) = \rho = N/V.$$

In an ideal gas, different particles are uncorrelated. As a result, for an ideal gas, the joint two-particle distribution, $P^{(2/N)}(\mathbf{r}_1, \mathbf{r}_2)$, factors as $P^{(1/N)}(\mathbf{r}_1)P^{(1/N)}(\mathbf{r}_2)$. Thus, for an ideal gas

$$\rho^{(2/N)}(\mathbf{r}_1, \mathbf{r}_2) = \frac{N(N-1)}{V^2} = \rho^2(1 - N^{-1}) \approx \rho^2,$$

where the last equality neglects the difference between $N-1$ and N. [There are situations (not treated in this text) for which the subtle distinction between N and $N-1$ actually becomes important.] In view of the ideal gas result for $\rho^{(2/N)}(\mathbf{r}_1, \mathbf{r}_2)$, it seems appropriate to

[*] If a fluid has a density $\rho = N/V$, the average number of particles in a microscopic volume Ω is $\rho\Omega$. (Suppose $\Omega = 1 \text{ Å}^3$. Then $\rho\Omega \sim 10^{-2}$ in a liquid and 10^{-4} in a gas.) The probability that specifically particle 1 is in that volume is $N^{-1}\rho\Omega \sim 10^{-23}\rho\Omega$.

introduce

$$g(\mathbf{r}_1, \mathbf{r}_2) = \rho^{(2/N)}(\mathbf{r}_1, \mathbf{r}_2)/\rho^2$$

or

$$h(\mathbf{r}_1, \mathbf{r}_2) = [\rho^{(2/N)}(\mathbf{r}_1, \mathbf{r}_2) - \rho^2]/\rho^2$$
$$= g(\mathbf{r}_1, \mathbf{r}_2) - 1,$$

which is the fractional deviation from the ideal gas approximation to the true two-particle distribution function. For an isotropic fluid, these functions depend upon $|\mathbf{r}_1 - \mathbf{r}_2| = r$ only; that is,

$$g(\mathbf{r}_1, \mathbf{r}_2) = g(r),$$
$$h(\mathbf{r}_1, \mathbf{r}_2) = h(r) = g(r) - 1.$$

The quantity $g(r)$ is called a *radial distribution function*. It is also often referred to as a pair correlation function or pair distribution function. The quantity $h(r)$ is called the pair correlation function too.

As already noted, $\rho^{(1/N)}(\mathbf{r}_1) = \rho$ for a uniform system. Thus

$$\rho^{(2/N)}(0, \mathbf{r})/\rho = \rho g(r)$$
$$= \text{conditional probability density that a}$$
$$\text{particle will be found at } \mathbf{r} \text{ given that}$$
$$\text{another is at the origin.}$$

The reasoning behind this result is based upon a theorem of probability statistics: If x and y are random variables with a joint distribution $P(x, y)$, then the conditional probability distribution for y given a specific value of x is $P(x, y)/p(x)$, where $p(x)$ is the probability distribution for x. Alternatively,

$$\rho g(r) = \text{average density of particles at } \mathbf{r} \text{ given that}$$
$$\text{a tagged particle is at the origin.}$$

When the terminology "liquid structure" is used, one is referring to quantities like $g(r)$. Unlike a crystal, the single particle distribution for a fluid is trivial. It is simply a bulk property, the density. The isotropic symmetry must be broken (e.g., by stating that a particle is known to be at a particular location). Once the symmetry is broken, there is interesting microscopic structure. Thus, for a fluid, one thinks about *relative* arrangements of atoms or molecules rather than absolute arrangements. To get a feeling for what pair correlation functions look like, let's consider a simple atomic liquid like argon. A schematic view of the liquid (drawn in two dimensions for artistic convenience) is shown in Fig. 7.1. In this picture, σ is the van der

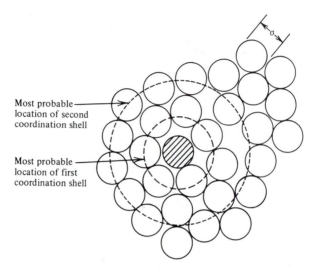

Fig. 7.1. Simple liquid structure.

Waals diameter* (about 3.4 Å for argon); the cross-hatched atom is the one we will take to be at the origin. The atoms are drawn close together because of typical liquid densities, $\rho\sigma^3 \sim 1$. Since the fluid is dense, there is a strong likelihood that a first neighbor shell will be found around $r = \sigma$. The nearest neighbors, which comprise the *first coordination shell,* tend to exclude the next nearest neighbors from an intermediate region around $r \approx (3/2)\sigma$. Thus, $g(r)$ will be less than unity in that region, and it will peak above the uncorrelated result near $r = 2\sigma$. Indeed, Fig. 7.2 shows what $g(r)$ for a simple atomic liquid looks like. The second peak corresponds to the most probable location for the next nearest neighbors. These neighbors comprise the *second coordination shell.* This layering manifests the granularity (non-continuum nature) of the liquid. It shows up in an oscillatory form for $g(r)$, which persists until r is larger than the range of correlations (typically a few molecular diameters in a dense liquid). In the dilute gas phase, the range of correlations is just the range of the intermolecular pair potential and there is no layering. [We will return to the gas phase $g(r)$ later at which point we will derive why it looks as we have drawn it in the figure.]

Notice both in the picture of the liquid and in the graph of $g(r)$ that there is a finite density of particles even in "unlikely" regions like $r = (3/2)\sigma$. This is one of the features that distinguishes a liquid

---

* Defined roughly as the distance of closest approach between two atoms during a physical (non-chemical) encounter.

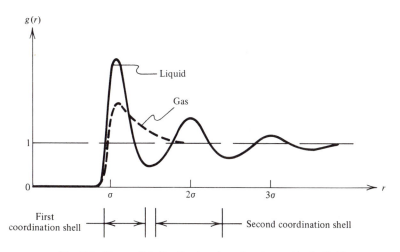

Fig. 7.2. The radial distribution function for a simple fluid.

from a crystalline solid. Without it, the possibility of diffusion would be drastically reduced.

A schematic view of a solid (once again drawn in two dimensions) is shown in Fig. 7.3. A radial distribution function for a three-dimensional (fcc or bcc) low temperature solid is depicted in Fig. 7.4. This function is the angle average of $g(r)$. Notice that the ordering of the first coordination shell in a solid allows the second nearest neighbors to be located at a distance $\sqrt{2}\,\sigma$ (or $\sqrt{3}\,\sigma$ in two dimensions) from the tagged atom. This decrease from 2σ accounts for the fact that for simple systems, the *bulk* density of the solid phase is higher than that for the liquid.

A quantitative comparison of the $g(r)$'s for liquid and solid argon at its triple point (which corresponds to a low temperature liquid and a high temperature crystal, respectively) is shown in Fig. 7.5.

The reader might wonder if the density of nearest neighbors is also greater in the solid than the liquid. The number of neighbors within a

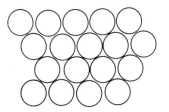

Fig. 7.3. Two-dimensional crystalline array of spherical particles.

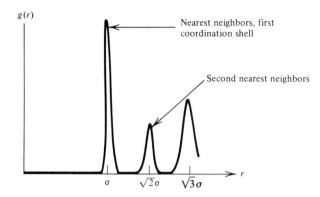

Fig. 7.4. Radial distribution function for a highly ordered solid.

distance r from a central atom is

$$n(r) = 4\pi\rho \int_0^r x^2 g(x)\, dx.$$

Exercise 7.8 Justify this formula.

When the integration is over the first coordination shells, the formula yields

$$n(\text{first coordinate shell}) \approx 12$$

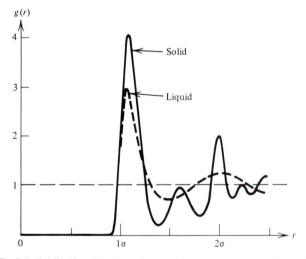

Fig. 7.5. Radial distribution functions for liquid and solid argon at the triple point $(\sigma = 3.4\ \text{Å})$.

for both the solid and the liquid. (This is the result in three dimensions. What would it be in two?) Further, it is usually found that $g(r)$ for a solid peaks at a slightly larger distance than that for the liquid. Thus, if these criteria are used, the nearest-neighbor density of a liquid is not less than the nearest-neighbor density of a solid. The difference between the two phases comes from the ordering of the first coordination shell. This ordering allows the closer approach of the second coordination shell. It also prohibits appreciable concentrations of particles between the first two coordination shells. This behavior gives rise to the long range order present in a solid (and absent in a liquid), and it severely inhibits diffusion.

7.3 Reversible Work Theorem

The reduced distribution functions are related to a Helmholtz free energy by a remarkable theorem:

$$g(r) = e^{-\beta w(r)},$$

where $w(r)$ is the reversible work for a process in which two tagged particles are moved through the system from infinite separation to a relative separation r. Clearly,

$$w(r) = w(r; \beta, \rho).$$

Since the process is performed reversibly at constant N, V, and T, $w(r)$ is the change in Helmholtz free energy for the process.

To prove this theorem, we consider the solvent averaged force between a pair of particles, say 1 and 2. Here, "solvent" refers to all the particles in the system except those that are tagged. By performing the average over all configurations with particles 1 and 2 held fixed at $\mathbf{r}_1$ and $\mathbf{r}_2$, respectively, we find that the averaged force is

$$-\left\langle \frac{d}{d\mathbf{r}_1} U(r^N) \right\rangle_{\mathbf{r}_1, \mathbf{r}_2 \text{ fixed}} = \frac{-\int d\mathbf{r}_3 \cdots d\mathbf{r}_N (dU/d\mathbf{r}_1) e^{-\beta U}}{\int d\mathbf{r}_3 \cdots d\mathbf{r}_N e^{-\beta U}}$$

$$= +k_B T \left[\frac{d}{d\mathbf{r}_1} \int d\mathbf{r}_3 \cdots d\mathbf{r}_N e^{-\beta U} \right] \Big/ \int d\mathbf{r}_3 \cdots d\mathbf{r}_N e^{-\beta U}$$

$$= k_B T \frac{d}{d\mathbf{r}_1} \ln \int d\mathbf{r}_3 \cdots d\mathbf{r}_N e^{-\beta U} = k_B T \frac{d}{d\mathbf{r}_1} \ln \left[N(N-1) \right.$$

$$\left. \times \int d\mathbf{r}_3 \cdots d\mathbf{r}_N e^{-\beta U} \Big/ \int d\mathbf{r}^N e^{-\beta U} \right] = k_B T \frac{d}{d\mathbf{r}_1} \ln g(\mathbf{r}_1, \mathbf{r}_2).$$

This result shows that $-k_B T \ln g(|\mathbf{r}_1 - \mathbf{r}_2|)$ is a function whose gradient gives the force between particles 1 and 2 averaged over the equilibrium distribution for all the other particles. Integration of the averaged force yields the reversible work. Thus, $w(r) = -k_B T \ln g(r)$ is indeed the reversible work as described above. As the derivation of this result might suggest, $w(r)$ is often called a *potential of mean force*.

7.4 Thermodynamic Properties from $g(r)$

Up to this point, we have not specified a form for $U(r^N)$. The simplest possibility is

$$U(r^N) = \sum_{i>j=1}^{N} u(|\mathbf{r}_i - \mathbf{r}_j|),$$

where $u(r)$ is a pair potential, as sketched in Fig. 7.6, where we have taken $u(\infty)$ as the zero of energy. The pair decomposable form for $U(r^N)$ is only an approximation even for atoms. The reason is that the internal structure of an atom involves fluctuating charge distributions (the quantal electrons). The potential energy expressed as only a function of nuclear coordinates arises in some fashion by integrating out the intra-atomic charge fluctuations. If the fluctuations are large in size, the resulting energy function will be complicated, coupling together more than pairs of particles. For most atoms, however, the charge fluctuations are relatively small, and pair decomposability is a good approximation. A commonly used two-parameter expression for the pair potential, $u(r)$, is the Lennard–Jones potential,

$$u(r) = 4\varepsilon[(\sigma/r)^{12} - (\sigma/r)^6].$$

Exercise 7.9 Show that the minimum in the Lennard–Jones potential is located at $r_0 = 2^{1/6}\sigma$ and that $u(r_0) = -\varepsilon$.

For interparticle separations larger than r_0, $u(r)$ is attractive, and for smaller separations, $u(r)$ is repulsive. The attractive interactions are due to the dipole-dipole coupling of the charge fluctuations in the separated atoms. Here, note that it is the *average* dipole of an isolated atom that is zero. Instantaneous fluctuations can have nonspherical symmetry, and the resulting dipole in one atom can

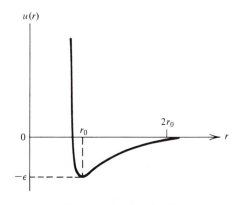

Fig. 7.6. A pair potential.

couple to that in another leading to attractive interactions. Such interactions are called London dispersion potentials, and these potentials are proportional to r^{-6}. There is quite a different origin to the repulsive forces between atoms at small separations. At small r, the electronic clouds on the two atoms must distort to avoid spatial overlap excluded by the Pauli principle. By distorting the electron distributions, the energy of the atoms increases thus leading to a repulsion between atoms.

The Lennard–Jones potential has an attractive branch that is asymptotic to r^{-6}, but the particular 6–12 algebraic form is not fundamental. The most important features of the Lennard–Jones pair potential have to do with length scales. In particular, r_0 is roughly 10% larger than σ, the distance at which $u(r) = 0$. Further, the range of $u(r)$ is roughly $2r_0$. This scaling holds empirically for most noble gas atoms and to a lesser extent for some simple molecules like N_2 and O_2.

With the pair decomposable approximation, we can calculate the internal energy as

$$\langle E \rangle = \langle K(p^N) \rangle + \langle U(r^N) \rangle$$
$$= N \langle p^2/2m \rangle + \left\langle \sum_{i>j=1}^{N} u(|\mathbf{r}_i - \mathbf{r}_j|) \right\rangle.$$

The first term on the right is $(3/2)Nk_BT$. The second term is a sum of $N(N-1)/2$ equivalent contributions (the number of separate pairs).

Each one has the value $\langle u(r_{12}) \rangle$, where $r_{12} = |\mathbf{r}_1 - \mathbf{r}_2|$. Thus

$$\left\langle \sum_{i>j=1}^{N} u(r_{ij}) \right\rangle = \frac{1}{2} N(N-1) \langle u(r_{12}) \rangle$$

$$= \frac{1}{2} \frac{N(N-1) \int dr^N u(r_{12}) \exp\left[-\beta U(r^N)\right]}{\int dr^N \exp\left[-\beta U(r^N)\right]}$$

$$= \frac{1}{2} \int d\mathbf{r}_1 \int d\mathbf{r}_2 \, u(r_{12}) N(N-1)$$

$$\times \int dr^{N-2} e^{-\beta U(r^N)} \Big/ \int dr^N e^{-\beta U}$$

$$= \frac{1}{2} \int d\mathbf{r}_1 \int d\mathbf{r}_2 \, \rho^{(2/N)}(\mathbf{r}_1, \mathbf{r}_2) u(r_{12}).$$

For a uniform system, $\rho^{(2/N)}(\mathbf{r}_1, \mathbf{r}_2) = \rho^2 g(r_{12})$. Thus, it is convenient to change variables of integration from $(\mathbf{r}_1, \mathbf{r}_2)$ to $(\mathbf{r}_{12}, \mathbf{r}_1)$. The $\mathbf{r}_1$ integration can be done freely to give a volume V. Thus,

$$\left\langle \sum_{i>j=1}^{N} u(r_{ij}) \right\rangle = (V\rho^2/2) \int d\mathbf{r} \, g(r) u(r)$$

$$= \frac{1}{2} N \int d\mathbf{r} \, \rho g(r) u(r).$$

We can understand this result on physical grounds. For each particle, there are $4\pi r^2 \rho g(r) \, dr$ neighbors in a shell of radius r and thickness dr, and the energy of interaction between the central particles and these neighbors is $u(r)$. The factor of $1/2$ is a symmetry number that corrects for double counting.

Combining these results yields

$$\langle E \rangle / N = \frac{3}{2} k_B T + \frac{1}{2} \rho \int d\mathbf{r} \, g(r) u(r).$$

Exercise 7.10 Express the internal energy in terms of reduced distribution functions when

$$U(r^N) = \sum_{i>j=1} u(r_{ij}) + \sum_{i>j>l=1} u^{(3)}(\mathbf{r}_i - \mathbf{r}_j, \mathbf{r}_j - \mathbf{r}_l).$$

Exercise 7.11* Show that when $U(r^N)$ is pair decomposable, the pressure is given by

$$\beta p / \rho = 1 - (\beta \rho / 6) \int d\mathbf{r}\, g(r) r\, du(r) / dr.$$

This formula is called the *virial theorem* equation of state. [*Hint*: In the configurational partition, for $1 \le i \le N$, change coordinates to $\mathbf{x}_i = V^{-1/3} \mathbf{r}_i$, so that $d\mathbf{x}_i = V^{-1} d\mathbf{r}_i$, and the limits of integration no longer depend upon volume.]

To see how the formulas for thermodynamic properties work, we need a theory for $g(r)$. One way to estimate $g(r)$ focuses attention on the potential of mean force, $w(r)$. We can separate $w(r)$ into two parts:

$$w(r) = u(r) + \Delta w(r).$$

The pair potential, $u(r)$, describes the reversible work to move the particles in a vacuum. Thus, $\Delta w(r)$ is the contribution to $w(r)$ due to surrounding particles in the system. That is, $\Delta w(r)$ is the change in Helmholtz free energy of the solvent due to moving particles 1 and 2 from $|\mathbf{r}_1 - \mathbf{r}_2| = \infty$ to $|\mathbf{r}_1 - \mathbf{r}_2| = r$. Clearly, in the low density limit

$$\lim_{\rho \to 0} \Delta w(r) = 0.$$

As a result,

$$g(r) = e^{-\beta u(r)}[1 + O(\rho)].$$

For higher densities, one must grapple with the deviations of $\Delta w(r)$ from zero. In the most successful approaches, one estimates $\Delta w(r)$ in terms of $\rho g(r)$ and $u(r)$. These approaches yield integral equations for $g(r)$ that are essentially mean field theories. We will not discuss these more advanced treatments here. Instead, we will consider the low density limit.

From the energy equation we have

$$\Delta E / N = (\rho / 2) \int d\mathbf{r}\, g(r) u(r)$$

$$= (\rho / 2) \int d\mathbf{r}\, e^{-\beta u(r)} u(r)[1 + O(\rho)],$$

where we have used the low density result for $g(r)$, and ΔE is defined as $E - E_{\text{ideal}}$. Note that

$$\Delta E / N = \partial(\beta \Delta A / N) / \partial \beta,$$

where ΔA is the excess (relative to the ideal gas) Helmholtz free energy, that is,

$$-\beta\Delta A = \ln\left(Q/Q_{\text{ideal}}\right).$$

Thus, integrating the molecular expression with respect to β yields

$$-\beta\Delta A/N = (\rho/2)\int d\mathbf{r}f(r) + O(\rho^2),$$

where

$$f(r) = e^{-\beta u(r)} - 1.$$

From this expression for the free energy we can obtain the pressure p via

$$\rho^2\frac{\partial(\beta\Delta A/N)}{\partial\rho} = \beta p - \rho.$$

The differentiation yields

$$\beta p = \rho + \rho^2 B_2(T) + O(\rho^3),$$

where

$$B_2(T) = -\frac{1}{2}\int d\mathbf{r}f(r)$$

is called the *second virial coefficient*.

Exercise 7.12 Insert $g(r) \approx \exp\left[-\beta u(r)\right]$ into the virial theorem and show that the same equation is obtained for the second virial coefficient. [*Hint:* You will need to integrate by parts.]

Exercise 7.13 Evaluate the second virial coefficient for a hard sphere system,

$$u(r) = \infty, \quad r < \sigma,$$
$$= 0, \quad r > \sigma,$$

and for a square well system

$$u(r) = \infty, \quad r < \sigma,$$
$$= -\varepsilon, \quad \sigma < r < \sigma',$$
$$= 0, \quad r > \sigma'.$$

Estimate the Boyle temperature, T_B [the temperature at which $B_2(T)$ is zero].

Exercise 7.14 Sketch a graph of $B_2(T)$ for a Lennard–Jones potential.

7.5 Measurement of $g(r)$ by Diffraction

Let us now consider how pair correlation functions can be measured. The measurement will have to probe distances of the order of or smaller than Angstroms. Thus, if radiation is used, its wavelength must be smaller than 1 Å; such wavelengths are obtained with X-rays or neutrons. The elementary theory of X-ray scattering is similar to that for neutrons. We treat X-rays here.

A schematic view of an X-ray scattering experiment is shown in Fig. 7.7. The scattered wave at the detector due to scattering from one atom at $\mathbf{R}_s$ is

$$\begin{bmatrix} \text{atomic scattering} \\ \text{factor} \end{bmatrix} |\mathbf{R}_D - \mathbf{R}_s|^{-1} \exp \{i[\mathbf{k}_{\text{in}} \cdot \mathbf{R}_s + \mathbf{k}_{\text{out}} \cdot (\mathbf{R}_D - \mathbf{R}_s)]\}.$$

(This is the spherical wave, first Born approximation.) If the detector is far from the scattering center,

$$|\mathbf{R}_D - \mathbf{R}_s| \approx |\mathbf{R}_D - \mathbf{R}_c|,$$

where $\mathbf{R}_c$ is the center of the scattering cell. For that case, the scattered wave at the detector is

$$f(k) |\mathbf{R}_D - \mathbf{R}_c|^{-1} e^{i\mathbf{k}_{\text{out}} \cdot \mathbf{R}_D} e^{-i\mathbf{k} \cdot \mathbf{R}_s},$$

where

$$\mathbf{k} = \mathbf{k}_{\text{out}} - \mathbf{k}_{\text{in}}$$

is the momentum transfer (within a factor of $\hbar$) for the scattered X-ray, and $f(k)$ is the atomic scattering factor. (It depends upon k. Why?) Now consider the vector diagram in Fig. 7.8. Since photons scatter nearly elastically, $|\mathbf{k}_{\text{in}}| \approx |\mathbf{k}_{\text{out}}|$. As a result,

$$k = |\mathbf{k}| = (4\pi/\lambda_{\text{in}}) \sin (\theta/2).$$

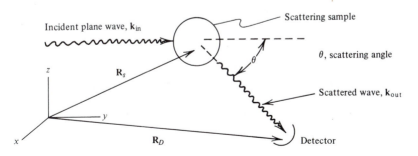

Fig. 7.7. X-ray scattering.

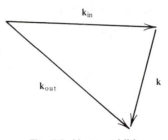

Fig. 7.8. Vector addition.

Exercise 7.15 Derive this formula for elastic scattering.

Since each atom in the system scatters, we have a superposition of waves at the detector:

$$\text{(total scattered wave)} = f(k) \frac{e^{i\mathbf{k}_{\text{out}} \cdot \mathbf{R}_D}}{|\mathbf{R}_c - \mathbf{R}_D|} \sum_{j=1}^{N} e^{-i\mathbf{k} \cdot \mathbf{r}_j},$$

where $\mathbf{r}_j$ is the position of the jth atom. The intensity is the square of the magnitude of the total wave, and the *observed* intensity is the ensemble average of that square:

$$I(\theta) = \text{observed intensity at detector}$$
$$= [|f(k)|^2/|\mathbf{R}_c - \mathbf{R}_D|^2]NS(k),$$

where

$$S(k) = N^{-1} \left\langle \sum_{l,j=1}^{N} \exp\left[i\mathbf{k} \cdot (\mathbf{r}_l - \mathbf{r}_j)\right] \right\rangle.$$

The quantity $S(k)$ is called the *structure factor*. It is related in a simple way to the Fourier transform of $g(r)$.

To see why, expand the sum over particles in $S(k)$ into self, $l = j$, and distinct, $l \neq j$, parts. There are N of the former and $N(N-1)$ of the latter. Thus,

$$S(k) = 1 + N^{-1}N(N-1)\langle e^{i\mathbf{k} \cdot (\mathbf{r}_1 - \mathbf{r}_2)}\rangle$$
$$= 1 + N^{-1} \frac{N(N-1) \int dr^N \, e^{i\mathbf{k} \cdot (\mathbf{r}_1 - \mathbf{r}_2)} e^{-\beta U}}{\int dr^N \, e^{-\beta U}}$$

$$= 1 + N^{-1} \underbrace{\int d\mathbf{r}_1 \int d\mathbf{r}_2 \; \rho^{(2)}(\mathbf{r}_1, \mathbf{r}_2)}_{\int d\mathbf{r}_1 \int d\mathbf{r}_{12} \; \rho^2 g(r_{12})} \; e^{i\mathbf{k} \cdot (\mathbf{r}_1 - \mathbf{r}_2)}$$

$$= 1 + \rho \int d\mathbf{r} \, g(r) e^{i\mathbf{k} \cdot \mathbf{r}}.$$

As a result, the measured structure factor determines the Fourier transform of $g(r)$. Since Fourier transforms are unique, $S(k)$ can be inverted to determine $g(r)$.

Exercise 7.16 Verify the algebraic details in this derivation and continue the reduction to show that

$$S(k) = 1 + (4\pi\rho/k) \int_0^\infty dr \sin{(kr)} r g(r).$$

7.6 Solvation and Chemical Equilibrium in Liquids

One of the most important aspects of liquid state science in the fields of biophysics and chemistry is the role of liquid environments in affecting conformational and chemical equilibria of solutes in solution. This is the subject of solvation, and here, too, reduced distribution functions are closely related to experimental observations of solvation.

To describe the relationship, we begin by deriving a formula for the chemical potential for a simple structureless solute species dissolved in a fluid at low solute concentrations. The total partition function is

$$Q = Q_S^{(\mathrm{id})} Q_A^{(\mathrm{id})} V^{-(N_A + N_S)} \int dr^{N_A} \int dr^{N_S}$$
$$\times \exp{[-\beta U_S(r^{N_S}) - \beta U_{AS}(r^{N_S}, r^{N_A})]},$$

where $Q_S^{(\mathrm{id})} Q_A^{(\mathrm{id})}$ is the ideal gas partition function for the solvent-solute mixture (it depends upon the numbers of solute and solvent molecules, N_A and N_S, respectively, the volume V, and temperature T), the potential energy U_S is the potential energy for the pure solvent (it is a function of the solvent configurations, r^{N_S}), and U_{AS} is the contribution to the potential energy due to the coupling between solvent and solute species. In this equation for the partition function we have left out a contribution to the potential energy due to the interactions between different solute species. These interactions are negligible at low enough solute contributions since in that case interactions between different solutes occur for only a negligible fraction of configuration space. The U_{AS} term, however, cannot be neglected since solutes are virtually always surrounded by and interacting with solvent species.

To analyze the effect of the U_{AS} term, we employ a trick known as the *coupling parameter method*. In particular, we let

$$Q_\lambda = Q_S^{(\mathrm{id})} Q_A^{(\mathrm{id})} V^{-(N_A+N_S)} \int dr^{N_A} \int dr^{N_S}$$

$$\times \exp\left[-\beta U_S(r^{N_S}) - \beta\lambda U_{AS}(r^{N_S}, r^{N_A})\right],$$

where $0 \le \lambda \le 1$ is the coupling parameter. When $\lambda = 0$, the solvent and solutes behave independently of one another, and when $\lambda = 1$, the Q_λ is the full partition function. Let us now consider the differential change in $\ln Q_\lambda$ due to changing λ. Within a factor of $-\beta$, $\ln Q_\lambda$ is the Helmholtz free energy for a system with total potential energy $U_S + \lambda U_{AS}$. By studying the change of $\ln Q_\lambda$ with respect to changing λ, we are therefore studying the reversible work to change the solvent–solute coupling. In view of the previous formula,

$$\frac{d\ln Q_\lambda}{d\lambda} = \frac{\int dr^{N_A} \int dr^{N_S}(-\beta U_{AS}) \exp\left(-\beta U_S - \beta\lambda U_{AS}\right)}{\int dr^{N_A} \int dr^{N_S} \exp\left(-\beta U_S - \beta\lambda U_{AS}\right)},$$

where we have omitted the arguments of the potential energy functions for notational convenience.

To proceed, we need to say something more about $U_{AS}(r^{N_S}, r^{N_A})$. We will assume a pair decomposable form,

$$U_{AS}(r^{N_S}, r^{N_A}) = \sum_{i=1}^{N_A} \sum_{j=1}^{N_S} u_{AS}(|\mathbf{r}_{iA} - \mathbf{r}_{jS}|),$$

where $\mathbf{r}_{iA}$ is the position of the ith solute, and $\mathbf{r}_{jS}$ is the position of the jth solvent. Inserting this expression into the formula for $d\ln Q_\lambda/d\lambda$ and carrying out the same manipulations encountered in earlier sections yields

$$-k_B T d\ln Q_\lambda/d\lambda = N_A \int d\mathbf{r}\, u_{AS}(r)\rho_S g_{AS}(r; \lambda),$$

where $\rho_S = N_S/V$, and $g_{AS}(r; \lambda)$ is the radial distribution function for a solvent-solute pair when the total potential energy for the whole system is $U_S + \lambda U_{AS}$.

Exercise 7.17 Derive this result.

We can now formally integrate the derivative giving the free energy

$$A(N_S, N_A, V, T) = A_{id}(N_S, N_A, V, T) + \Delta A_S(N_S, V, T)$$
$$+ N_A \int_0^1 d\lambda \int d\mathbf{r}\, u_{AS}(r)\rho_s g_{AS}(r, \lambda),$$

where A_{id} is the ideal gas Helmholtz free energy of the solvent-solute mixture and ΔA_S is the excess (beyond the ideal gas) Helmholtz free energy for the pure solvent. The last equation was derived by noting that $\ln Q_0 = -\beta(A_{id} + \Delta A_S)$. Finally, to complete our analysis, we differentiate A with respect to N_A to obtain the chemical potential at infinite dilution:

$$\mu_A = \mu_A^{(id)} + \Delta\mu_A,$$

where

$$\Delta\mu_A = \int_0^1 d\lambda \int d\mathbf{r}\, \rho_s g_{AS}(r; \lambda) u_{AS}(r),$$

and $\mu_A^{(id)}$ is the chemical potential for species A in an ideal gas. In Exercise 7.32, you will develop a somewhat different derivation of this same result for $\Delta\mu_A$, and you will see that $\Delta\mu_A$ is measured experimentally by determining the Henry's Law constants for an ideal solution.

We now turn our attention to the situation where two solutes encounter one another in solution. The analysis of the statistics of this encounter leads us to the theory of chemical and conformational equilibria in solution. The analysis can be made quite general, and students are encouraged to try their hands at the generalizations. But, for the sake of simplicity, we confine the discussion to the chemical equilibrium

$$A + B \rightleftharpoons AB,$$

which may occur in the gas phase or in a liquid solvent. An example might be the dimerization of NO_2 to form N_2O_4 in a gaseous environment or in liquid CCl_4.

To apply the rules of classical statistical mechanics to this process, we must have a definition, preferably a configurational one, for when an AB species is formed. We will focus on the distance r between the centers of A and B, and say that an AB dimer is formed whenever $r < R$, where R is some length we must specify. We will take R to be the range of the covalent bonding energy, $u_{AB}(r)$, which would favor the formation of the dimer. Let

$$H_{AB}(r) = 1, \quad r < R,$$
$$= 0, \quad r \geq R.$$

Then the ratio of the classical intramolecular partition functions for the AB dimer and monomers in the gas phase is (neglecting any considerations of the internal structure of the A and B species)

$$q_{AB}^{(id)}/q_A^{(id)}q_B^{(id)} = (1/\sigma_{AB})\int d\mathbf{r}\, H_{AB}(r)e^{-\beta u_{AB}(r)},$$

where σ_{AB} is the symmetry number for the dimer (1 when $A \neq B$, and 2 when $A = B$), and the superscripts "id" indicate the expression is appropriate to a dilute ideal gas where intermolecular interactions are negligible. Accordingly, the equilibrium constant

$$K = \rho_{AB}/\rho_A\rho_B$$

is given in the gas phase by

$$K^{(id)} = (1/\sigma_{AB})\int d\mathbf{r}\, H_{AB}(r)e^{-\beta u_{AB}(r)}.$$

In a condensed phase, the liquid solvent plays a role in the free energetics for the association process. Consider carrying out the dimerization by moving an A and B pair reversibly through the solvent, starting with the pair separated by a macroscopic distance and eventually reaching a mutual separation of r. Except for requiring the solvent to remain at equilibrium, there is no restriction on the concentration of solute species in the solvent. Since the process is carried out reversibly, the solvent contribution to the change in Helmholtz free energy is $\Delta w_{AB}(r)$, the indirect part of the potential of mean force. The total free energetics is thus $u_{AB}(r) + \Delta w_{AB}(r)$. Hence, in a liquid

$$K = (1/\sigma_{AB})\int d\mathbf{r}\, H_{AB}(r)\exp\left[-\beta u_{AB}(r) - \beta\Delta w_{AB}(r)\right]$$

$$= K^{(id)}\int dr\, s_{AB}^{(id)}(r)y_{AB}(r),$$

where

$$s_{AB}^{(id)}(r) \propto H_{AB}(r)e^{-\beta u_{AB}(r)}$$

is the intermolecular distribution function for an AB dimer in the gas phase, and

$$y_{AB}(r) = e^{-\beta\Delta w_{AB}(r)}$$

is called a cavity distribution function. It is given that name since $y_{AB}(r)$ is the radial distribution function for a pair of hypothetical particles A and B that do not interact directly with each other and are dissolved at infinite dilution in the solvent. These hypothetical particles are therefore like cavities in the fluid.

Exercise 7.18 Show that in the liquid, the intramolecular distribution is given by

$$s_{AB}(r) = s_{AB}^{(id)}(r)y_{AB}(r) \bigg/ \int d\mathbf{r}\, s_{AB}^{(id)}(r)y_{AB}(r).$$

Exercise 7.19 Let $\Delta\mu_i$ denote the excess chemical potential (beyond that found in the gas phase) for species i in a liquid solvent. Show that

$$\Delta\mu_{AB} = \Delta\mu_A + \Delta\mu_B - k_B T \ln \int d\mathbf{r}\, s_{AB}^{(id)}(r)y_{AB}(r).$$

Exercise 7.20 From the solubility of saturated alkane chains in liquid water, is found that to an excellent approximation, $\Delta\mu$ for the normal isomer of C_nH_{2n+2} in water depends linearly upon n. Explain this observation. [*Hint*: Consider the excess chemical potential in terms of the reversible work to create certain arrangements of "cavity particles" in the fluid; and also note that the linearity with n is a good approximation but not an exact result.] For $n > 10$, it becomes difficult to measure $\Delta\mu$ because of the diminishing solubility of the alkanes in water. Nevertheless, if the $\Delta\mu$'s could be measured, do you think the linear dependence would persist for large n? Explain.

7.7 Molecular Liquids

When scattering experiments are performed on molecular (as opposed to atomic) fluids, one obtains a superposition of scattering from all pair separations of atoms both intramolecular and intermolecular. Thus, for a fluid where ρ is the number of *molecules* per unit volume, the experiments probe both

$$\rho g_{\alpha\gamma}(r) = \text{density of atoms } \gamma \text{ at position } \mathbf{r} \text{ given that an atom}$$
$$\alpha \text{ in } another \text{ molecule is at the origin}$$

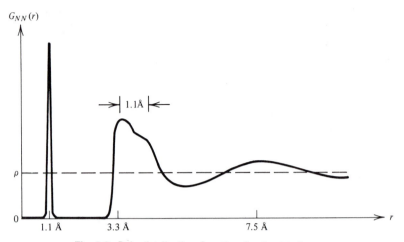

Fig. 7.9. Pair distribution function for liquid nitrogen.

and

$s_{\alpha\gamma}(r) =$ probability distribution that atom γ is at position
r given that another atom α in the *same* molecule
is at the origin.

Diffraction experiments determine a linear combination (in Fourier transform space) of

$$G_{\alpha\gamma}(r) = s_{\alpha\gamma}(r) + \rho g_{\alpha\gamma}(r).$$

For liquid nitrogen, the pair distribution is sketched in Fig. 7.9. The sharp peak at 1.1 Å is due to the intramolecular structure. In particular, the N–N bond length for an N_2 molecule is $L = 1.1$ Å. The remaining features are interpreted as follows: Since liquids are dense, it is more likely that molecules will be touching their neighbors. Thus, the main peak at 3.3 Å indicates that the van der Waals diameter of a nitrogen atom is roughly $\sigma = 3.3$ Å. Since each atom is attached to another via a chemical bond, and since each atom is touching atoms of neighboring molecules, it is likely that a tagged atom will also have neighboring atoms at $\sigma + L$. This is the reason for the shoulder or auxiliary peak found in $G_{NN}(r)$ at $r \approx (3.3 + 1.1)$Å.

Integration over the first coordination shell gives[*]

$$n = 4\pi \int_{3\,\text{Å}}^{5.6\,\text{Å}} G_{NN}(r)r^2\,dr$$

$$\approx 12.$$

[*] Since each N_2 molecule has two atoms, the total density of N atoms around a central atom is $2\rho G_{NN}(r)$. Hence, the total number of atoms in the first coordination shell is roughly 24.

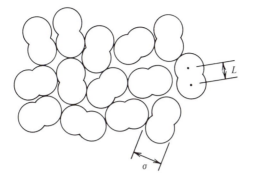

Fig. 7.10. Liquid structure with diatomic molecules.

Thus, each N_2 molecule has roughly 12 neighbors in the first coordination shell. This suggests that the structure in that shell after averaging over each of the particles is somewhat like that in a simple atomic fluid. Indeed, the location of the second coordination shell, $7.5 \text{ Å} \approx 2(\sigma + L/2)$, is in agreement with this idea. But notice that the oscillations in $G_{NN}(r)$ die out in a shorter distance than that for a liquid composed of spherical particles. Also, the peaks are lower and broader for the diatomic fluid. The reason is due to the presence of two length scales, σ and L, rather than just the van der Waals diameter σ. The second length introduces a larger variety of possibilities for the local intermolecular structure, and this variety produces a randomization that washes out the pair correlations.

A schematic picture of a region of the liquid we have just described is shown in Fig. 7.10.

Exercise 7.21 The carbon-carbon pair distribution function for *n*-butane, $CH_3CH_2CH_2CH_3$, is sketched in Fig. 7.11. Explain the qualitative features seen in this curve. (You will need to note that *n*-butane has three stable conformational states: *trans, gauche+,* and *gauche−*.)

Liquids nitrogen and butane are *nonassociated* liquids. That is, their intermolecular structure can be understood in terms of packing. There are no highly specific intermolecular attractions in these systems. Perhaps the most important example of an *associated* liquid is water. Here, the linear hydrogen bonding tends to produce a local tetrahedral ordering that is distinct from what would be predicted by only considering the size and shape of the molecule. The strength of a linear hydrogen bond is roughly $10k_B T$ at room temperature. This

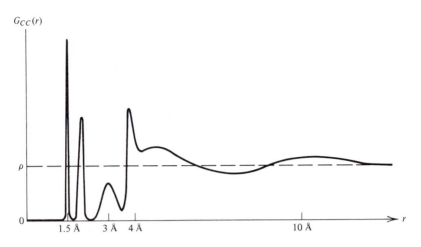

Fig. 7.11. Pair distribution function for liquid *n*-butane.

should be compared with the typical size of attractive interactions in nonassociated liquids: 1 to $2k_BT$ at the triple point.

A heuristic picture of three water molecules with the outer two hydrogen bonding with the one in the middle is shown in Fig. 7.12. A linear hydrogen bond is formed when a proton overlaps with or gets close to one of the fictitious lone pair electron lobes (or "orbitals"). With the three molecules shown, it is clear why the hydrogen bonding favors tetrahedral ordering. Of course, this ordering persists over large distances only in the ice phases. In the liquid, there is sufficient randomness even in the first coordination shell that the

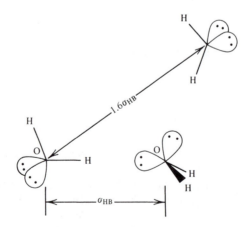

Fig. 7.12. Three water molecules and two hydrogen bonds.

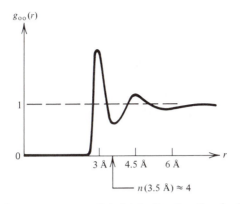

Fig. 7.13. Oxygen-oxygen radial distribution function for liquid water.

tetrahedral ordering is rapidly lost within one or two molecular diameters. In fact, the oxygen-oxygen pair distribution function for water at room temperature looks as shown in Fig. 7.13. The first peak is located at roughly $2.8 \text{ Å} = \sigma_{HB}$, the hydrogen bond distance. The second peak is found at $4.5 \text{ Å} \approx 1.6 \times 2.8 \text{ Å}$, just as would be expected from a tetrahedral structure. Further, when integrated over the first coordination shell,

$$n = 4\pi\rho \int_0^{3.5 \text{ Å}} g_{OO}(r)r^2 \, dr \approx 4.$$

But beyond the second coordination shell, the pair correlations essentially vanish. The orientational correlations vanish in even a shorter distance than the translational correlations. This fact can be established from the behavior of the OH and HH intermolecular distribution functions which are shown in Fig. 7.14. It is clear that all the features in these curves arise from the structure of the first

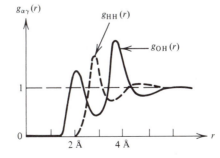

Fig. 7.14. H–H and O–H distributions in liquid water.

coordination shell only. Thus, the structure of liquid water is a highly distorted random network of hydrogen bonded species.

It is interesting to juxtapose the distribution functions for water with those for a nonassociated fluid. Recall that the second peak in $g(r)$ for a simple fluid occurs at twice the position of the first peak. There is no remnant of the second neighbor peak for the solid located at $\sqrt{2}\,\sigma$. Further, efficient packing of the nonassociated liquid tends to make each molecule have 12 neighbors in its first coordination shell. In contrast, a water molecule in water has roughly four nearest neighbors, and there is a clear remnant of the ice structures in which the second neighbors appear at $1.6\sigma_{HB}$. This local water structure is tenuous. The tetrahedral ordering is an inefficient means of filling space, since a large fraction of space remains unoccupied. The large hydrogen bond energy is required to promote this local structure which is so unfavorable as far as packing is concerned. But it is just barely able to compete with the packing. As a result, the structure is fragile as manifested by its unusually large temperature dependence.

Exercise 7.22* One anomalous behavior of water is its unusually large heat capacity, C_v. For water, C_v(liquid) $-$ C_v(gas) $\approx 20R$, whereas for most liquids the number is 2 to $5R$. Show that this behavior is a direct manifestation of the relatively large temperature dependence in the water structure. [*Hint*: Think about relating the internal energy to pair correlation functions.]

Since tetrahedral ordering leaves empty space in the fluid, it is clear that the isothermal application of pressure (which will decrease the volume per particle) will tend to rupture the local structure. Thus, the ordering in water will *decrease* with increasing pressure. In nonassociated liquids, higher pressure leads to greater packing and thus more order. The fact that increasing pressure or density can lead to less order in water is directly responsible for the density maximum in water found, for example, at 4°C and 1 atm pressure.

Exercise 7.23 Verify this statement. [*Hint*: Note that $(\partial s/\partial p)_T = -(\partial v/\partial T)_p$.]

7.8 Monte Carlo for Hard Disks

In the preceding sections, we have described many of the general properties of liquid phase pair correlation functions. Further, we have shown how the qualitative behavior of these functions can be anticipated from a rudimentary knowledge of the interparticle potentials acting between atoms and molecules. For instance, packing effects or geometrical considerations that focus on the shapes of molecules can be invoked to understand the structural behavior of most dense fluids. However, for associated liquids such as water, one must consider the effects of the highly directional and very strong hydrogen bonds.

To go beyond the qualitative pictures and develop a quantitative treatment of the connection between these potentials of interaction and liquid structure requires calculations that in some way accurately sample the multiparticle distribution

$$P(r^N) \propto \exp[-\beta U(r^N)].$$

There are several approaches to this problem. Some of them utilize analytical treatments that involve perturbation theories and mean field approximations of various sorts. A few such techniques have been remarkably successful in explaining the nature of the liquid state. Despite their intrinsic computational simplicity, however, the analytical methods cannot be properly developed and tested without the simultaneous implementation of computer simulations. The simulations are numerical experiments used to test the accuracy of the approximations used in the analytical theories. Further, while numerically tedious, simulations are conceptually far simpler than the analytical methods, the latter requiring much greater mathematical sophistication than the former.

In Chapter 6 we introduced the Monte Carlo technique as one such method for studying the fluctuations in discretized lattice models. In this section we show how this numerical sampling scheme can be extended to fluids where the degrees of freedom are continuous. The particular model we consider is a two-dimensional fluid of hard disks. This model is perhaps the simplest system which exhibits many of the most important structural phenomena that occur with dense fluid systems found in nature.

Exercise 7.24 With computer simulations, scientists have observed that a two-dimensional fluid of hard disks, each of diameter σ, seems to become unstable when it is compressed to a density higher than 70% of the closest packed density, ρ_{CP}. Above this density, the system freezes into a periodic crystal. What is the value of ρ_{CP}? What would you predict as the crystal structure of the hard disk solid?

The generalization from Metropolis Monte Carlo for Ising models to Metropolis Monte Carlo for fluids is relatively straightforward. For the continuous variables that characterize the configurations of a fluid, however, the numerical arithmetic is more difficult and requires greater computation time than that for an Ising model. In the latter case, most of the arithmetic operations can be reduced to multiplying zeros and ones.

This additional *numerical* complexity might be contrasted with the added *conceptual* complexity that is needed to generalize the perturbation theory and mean field treatments of Ising models presented in Chapter 5 to analogous theories for continuous fluids.

Exercise 7.25* Try to generalize the molecular mean field treatment of Sec. 5.4 to the case of a continuous fluid with pair interactions $u(r)$. If you succeed, the analog of the transcendental mean field equation $m = \tanh[\beta\mu H + \beta z J m]$ is

$$\langle \rho(\mathbf{r}) \rangle = c \exp\left[-\beta\phi(\mathbf{r}) + \beta \int d\mathbf{r}' \, \langle \rho(\mathbf{r}') \rangle u(|\mathbf{r} - \mathbf{r}'|) \right],$$

where c is a constant of proportionality, and $\phi(\mathbf{r})$ is an external potential energy field. Notice that this relationship is an *integral equation* for $\langle \rho(\mathbf{r}) \rangle$, the average density at position $\mathbf{r}$. Suggest methods (numerical and/or analytical) by which you might solve this equation.

The so-called "integral equation theories" of liquids are based upon mean field approximations like this one illustrated here.

The potential energy between a pair of particles ij in a hard disk fluid is

$$u(r_{ij}) = \infty, \quad r_{ij} < \sigma,$$
$$= 0, \quad r_{ij} > \sigma,$$

where

$$r_{ij}^2 = (x_i - x_j)^2 + (y_i - y_j)^2.$$

Here (x_i, y_i) are the Cartesian coordinates for particle i. In the program listed below, $N = 20$ disks are placed in a square cell of side length L. The particle density, $\rho = N/L^2$, is fixed by the value of L, and the disk diameter, σ, is the unit of length. Periodic boundary conditions are employed so that if a particle leaves the cell during a step of the simulation, another particle enters the cell at the other side. For example, if the center of particle i changes from (x_i, y_i) to

$(x_i + \delta, y_i)$ in a Monte Carlo step and if $x_i + \delta$ is a position outside the cell, then the new position for the center of particle i is actually taken as $(x_i + \delta - L, y_i)$. Notice that this type of boundary condition corresponds to simulating an infinite system by considering only those fluctuations which are periodically replicated from cell to cell.

After setting up an initial configuration of the system, $r^N = (x_1, y_1, \ldots, x_i, y_i, \ldots, x_N, y_N)$, a disk is identified by its position (x, y). Each disk is considered in turn. A possible new position for the disk under consideration is chosen by two random numbers, Δx and Δy, in the range $[-\text{del}, \text{del}]$, where "del" is adjusted experimentally to give about a 30% acceptance of the new position. (A "del" in the range $[0.05, 0.1]$ seems to work well in the program listed in the Appendix.) The possible new position is $(x + \Delta x, y + \Delta y)$. The energy difference, ΔU, between the possible new configuration, r'^N, and the old configuration, r^N, is either 0, if the disks do not overlap, or ∞, if any of the disks do overlap. Recall from Chapter 6 that according to the Metropolis algorithm for Monte Carlo, the new configuration will be accepted if

$$\exp(-\beta \Delta U) \geqslant x,$$

where x is a random number between 0 and 1. Otherwise, the move is rejected. In the case of hard disks, $\exp(-\beta U)$ is either 0 or 1. So the acceptance criteria is simply decided upon whether or not a new configuration introduces overlap of particles. The trajectory moves from step to step as described above, and the configuration at step $t + 1$ is

$$r^N(t + 1) = r^N(t)$$

if the attempted move to configuration r'^N caused an overlap of two disks, and

$$r^N(t + 1) = r'^N$$

if there was no overlap.

One property that can be computed from such a trajectory is the radial distribution function, $g(r)$. In particular, one may average the occurrence of particular pair separations. This average can be calculated by considering the average number of particles in a shell located a distance r from some tagged particle j. The thickness of the shell is taken to be 0.10. This increment of length will then be the smallest length over which we resolve $g(r)$. Let $\langle n_j(r) \rangle$ be the average number in the shell at distance r. Then

$$\langle n_j(r) \rangle = \frac{1}{T} \sum_{i=1}^{T} n_{ji}(r),$$

where $n_{ji}(r)$ is the number of particles in the shell on pass i. Here, T is the total number of passes, and one "pass" corresponds to N steps. This average is independent of which particle is the tagged particle. Let

$$\langle n(r) \rangle = \langle n_1(r) \rangle = \cdots = \langle n_N(r) \rangle,$$

or,

$$\langle n(t) \rangle = \frac{1}{N} \sum_{j=1}^{N} \langle n_j(r) \rangle$$

$$= \frac{1}{NT} \sum_{j=1}^{N} \sum_{i=1}^{T} n_{ji}(r).$$

If the particles were uncorrelated, the average number in the shell at distance r would be

$$\langle n(r) \rangle_{unc} = (\text{area of shell})\rho(N-1)/N,$$

where ρ is the density of the liquid and $(N-1)/N$ corrects for the inability of the tagged particle to be in the shell at distance r. Now $g(r)$ can be expressed as

$$g(r) = \frac{\langle n(r) \rangle}{\langle n(r) \rangle_{unc}}$$

$$= \frac{\sum_{j=1}^{N} \sum_{i=1}^{T} n_{ji}(r)}{(\text{area of shell})(N-1)T\rho}.$$

The program in the end-of-chapter Appendix uses this algorithm. It is written in BASIC for the Apple "Macintosh."

As the trajectory progresses, the statistics for $g(r)$ improve. Figure 7.15 taken from the Macintosh display illustrates the evolution of the statistics for the density $\rho\sigma^2 = 0.7$.

Exercise 7.26 Discuss how the accumulation of statistics for $g(r)$ depends on the size of the shell width.

Exercise 7.27 Experiment with the evolution of the Monte Carlo trajectory as one changes the size of "del" and thereby influences the percentage of accepted moves. Note that if "del" is too large, nearly all moves will be rejected, and the configurations of the system will not efficiently sample configuration space. Similarly, if "del" is too small, nearly all moves will be accepted, but the step size will be so small that sampling will still be inefficient. It would seem that there is an optimum midground. Can you think of a criterion for optimizing the choice of step size?

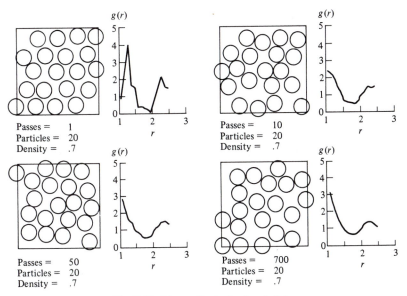

Fig. 7.15. Monte Carlo for hard disks.

Additional Exercises

7.28. Consider a dilute gas of argon atoms at a temperature T. Compute, as a function of T, the following mean values:

$$\text{(a) } \langle v_x^2 \rangle, \quad \text{(b) } \langle v_x^2 v_y^2 \rangle, \quad \text{(c) } \langle v^2 \rangle, \quad \text{(d) } \langle v_x \rangle,$$
$$\text{(e) } \langle (v_x + b v_y)^2 \rangle,$$

where v_x and v_y are Cartesian coordinates of the velocity **v** of one of the argon atoms. Discuss how your results will change if the gas is compressed isothermally to the point at which it liquifies, and it is further compressed to the point where it freezes.

7.29. Consider a system of N distinguishable non-interacting harmonic oscillators. The Hamiltonian is

$$\mathcal{H} = \sum_{i=1}^{N} p_i^2 / 2m + \sum_{i=1}^{N} \tfrac{1}{2} k \, |\mathbf{r}_i - \mathbf{r}_i^{(0)}|^2,$$

where $\mathbf{r}_i^{(0)}$ is the equilibrium position of the ith oscillator particle.

(a) Assume that the oscillators obey Schrödinger's equation

and determine the canonical partition function for this system.

(b) Assume that the oscillators obey Newton's equations of motion and determine the partition function for this system.

(c) Show that the results of (a) and (b) agree in the high temperature limit.

7.30. Consider a system of classical particles with both pairwise additive and three-body potentials. Show that the second virial coefficient is independent of the three-body potentials.

7.31. In this problem you will consider an equilibrium classical fluid of N hard rods confined to move on a line of length L as depicted in Fig. 7.16. The length of each individual rod is l, and $\rho_{CP} = l^{-1}$ is the close packed value of the density, $\rho = N/L$. The Helmholtz free energy for the system is given by

$$-\beta A = \ln\left[(N! \, \lambda^N)^{-1} \int_0^L dx_1 \cdots \int_0^L dx_N e^{-\beta U} \right],$$

where $\beta^{-1} = k_B T$, U is the total potential energy that depends upon the rod positions $x_1, \ldots, x_N$, and λ is the thermal DeBroglie wavelength. The pressure is given by

$$\beta p = \partial(-\beta A)/\partial L,$$

and the pair distribution function is defined by

$\rho g(x) =$ average density of rods at position x given that another is at the origin.

Note that x can be both positive and negative, and, for simplicity, assume the origin is far from the walls.

(a) At a high density (but less than ρ_{CP}), draw a labeled sketch of $g(x)$ for $x > 0$.

(b) Draw another labeled sketch of $g(x)$ for $x > 0$, but this time for the case in which $\rho \to 0$.

(c) Describe how $g(x)$ depends qualitatively upon temperature, T.

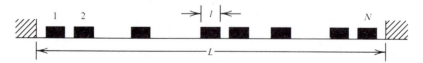

Fig. 7.16. Classical fluid of hard rods confined to a line of fixed length.

(d) For $\rho \to \rho_{CP}$, compute the value of the integral

$$\int_0^{(3/2)l} dx\, g(x).$$

(e) As a function of N, ρ, and T, determine the (i) average velocity of a rod, (ii) average speed of a rod, (iii) average kinetic energy of a rod, and (iv) internal energy of the full N particle system.

(f) With a (not too) tricky calculation, one can show that the density dependence of the pressure for this system is given by

$$\beta p = \rho/(1 - b\rho),$$

where b is independent of density. (i) Is b a function of temperature? (ii) Relate b to a second virial coefficient and use this connection to compute b in terms of β and l.

(g) Carry out the "tricky calculation" referred to in part (f) above.

7.32. (a) Consider gaseous argon. Neglect the internal structure of the argon atoms, and assume classical mechanics is valid. With a statistical thermodynamics calculation, show that in the low density limit, the chemical potential of gaseous argon is given by

$$\beta\mu = f(\beta) + \ln \rho,$$

where $f(\beta)$ is a function of $\beta = 1/k_B T$, only, and $\rho = N/V$ is the density of the argon.

(b) Now consider argon dissolved at low concentrations in liquid water. Show that the chemical potential for argon in this system is

$$\beta\mu = f(\beta) + \ln \rho + \beta\Delta\mu,$$

where $f(\beta)$ is the same quantity appearing in part (a), ρ is the density of argon in the water, and $\Delta\mu$ is an "excess" chemical potential that vanishes in the hypothetical limit in which argon and water molecules do not interact with one another (it is a function of the thermodynamic variables β and ρ_W, where ρ_W is the density of water).

(c) When the dilute argon vapor is in equilibrium with the argon-water solution, the pressure, p, of the argon vapor obeys Henry's Law

$$p = xk_H,$$

where x is the mole fraction of the argon in the water, that is,

$$x = \rho/(\rho + \rho_W) \approx \rho/\rho_W,$$

and k_H is the Henry's Law constant. Show that

$$k_H = \beta^{-1}\rho_W \exp(\beta\Delta\mu).$$

(d) The potential energy of the aqueous argon solution is very complicated depending upon the orientations and positions of all the water molecules and the coupling between each water molecule and each argon atom. Assume the form of the latter is such that

$$\sum_{i=1}^{N} u_{AW}(|\mathbf{r} - \mathbf{r}_i|)$$

is the potential energy associated with an argon atom at position $\mathbf{r}$. Here, $\mathbf{r}_i$ is the location of the center of the ith water molecule, and N is the total number of water molecules. This being the case, show that

$$\exp(-\beta\Delta\mu) = \left\langle \prod_{i=1}^{N} \exp[-\beta u_{AW}(|\mathbf{r} - \mathbf{r}_i|)] \right\rangle_W,$$

where $\langle \cdots \rangle_W$ indicates the canonical ensemble average over the coordinates of all the water molecules. (It is weighted by the Boltzmann factor with the total potential energy of all the water molecules.) [*Hint*: Conceive of $\Delta\mu$ as the difference between the Helmholtz free energy of water when the solute is present and that when the solute is absent.] Finally, use this result to show that

$$\Delta\mu = \int_0^1 d\lambda \int d\mathbf{r}\, \rho_W g_{AW}(r; \lambda) u_{AW}(r),$$

where $g_{AW}(r; \lambda)$ is the argon-water radial distribution function for one argon dissolved in water and this argon is coupled to the water via the pair potential $\lambda u_{AW}(r)$.

7.33.* Consider the virial theorem for the equation of state (see Exercise 7.11). For a *two*-dimensional fluid of hard disks, show that

$$\beta p/\rho = 1 - (\beta\rho/4)\int d\mathbf{r}\, g(r) r\, du(r)/dr$$

$$= 1 - (\beta\rho\pi/2)\int_0^\infty dr\, r^2 g(r)[du(r)/dr]$$

$$= 1 + (\rho\sigma^2\pi/2)g(\sigma^+),$$

where $g(\sigma^+)$ is the contact value of $g(r)$. {*Hint*: In the last equality, note that $-\beta g(r)[du(r)/dr] = y(r)d \exp[-\beta u(r)]/dr$, where $y(r) = \exp[-\beta \Delta w(r)]$ is the cavity distribution function. Furthermore, note that the derivative of a step function is a Dirac delta function.}

7.34. Use the Monte Carlo code for hard disks and the result of Exercise 7.33 to evaluate the pressure of the hard disk fluid. Note that to obtain $g(\sigma^+)$ you will need to extrapolate from r values greater than σ. Try a linear extrapolation. Compare your results with the estimates of the pressure given below. Comment on the sources of your errors (e.g., finite averaging time, small system size, etc.)

$(\rho_{CP}/\rho)^a$	$(\beta p/\rho)^b$
30	1.063
5	1.498
2	3.424
1.6	5.496
1.4	8.306

^{a} ρ_{CP} = closest packed density.
^{b} Source: J. J. Erpenbeck and M. Luban, *Phys. Rev. A* **32**, 2920 (1985). These numbers were obtained with very powerful computers. The uncertainty is in the third decimal place.

7.35. Consider a classical fluid with the potential energy

$$U(r^N) = \sum_{i>j=1}^{N} u(|\mathbf{r}_i - \mathbf{r}_j|).$$

Imagine dividing the pair potential into two parts,

$$u(r) = u_0(r) + u_1(r).$$

Call $u_0(r)$ the reference potential, and define the reference system as that system for which

$$\sum_{i>j=1}^{N} u_0(|\mathbf{r}_i - \mathbf{r}_j|)$$

is the total potential energy. The remaining part of the pair potential, $u_1(r)$, is then the perturbation pair potential.

(a) Show that the Helmholtz free energy of the full system is given by

$$A/N = A_0/N + \tfrac{1}{2}\rho \int_0^1 d\lambda \int d\mathbf{r}\, u_1(r) g_\lambda(r),$$

where A_0 is the reference system free energy and $g_\lambda(r)$ is the radial distribution for a hypothetical system which has the pair potential $u_0(r) + \lambda u_1(r)$. The reference system and the hypothetical system are at the same density $\rho = N/V$, temperature, and particle number N as the full system. [*Hint*: First compute $dA_\lambda/d\lambda$, where A_λ is the Helmholtz free energy for the hypothetical system, and then note

$$A - A_0 = \int_0^1 d\lambda\, (dA_\lambda/d\lambda).]$$

(b) Derive the following bound

$$A/N \leq A_0/N + \tfrac{1}{2}\rho \int d\mathbf{r}\, g_0(r) u_1(r),$$

where $g_0(r)$ is the radial distribution function for the reference system. [*Hint*: Recall the Gibbs-Bogoliubov-Feynman bound discussed in Chapter 5.]

Relations such as these form the basis of *thermodynamic perturbation theories* of fluids. These theories begin with the knowledge of the properties of a reference system fluid (e.g., properties of the hard sphere fluid determined from computer simulations) and then deduce the properties of other fluids by computing the change in free energy associated with changing the potential from the reference potential to those of interest.

Bibliography

The standard advanced treatise on the subject of this chapter is
  J. P. Hansen and I. R. McDonald, *Theory of Simple Liquids* (Academic Press, N.Y., 1976).
Other helpful textbook treatments are found in
  D. McQuarrie, *Statistical Mechanics* (Harper & Row, N.Y., 1976),
and
  H. L. Friedman, *A Course in Statistical Mechanics* (Prentice-Hall, Englewood Cliffs, N.J., 1985).
The structure of liquid water is considered in the review
  F. H. Stillinger, *Adv. Chem. Phys.* **31,** 1 (1975).
One of the central ideas in liquid state physics and chemistry is that for

non-associated liquids, the liquid structure is determined by packing effects or the shapes of molecules. This view is often referred to as the van der Waals picture of liquids, and it forms the basis for many successful theories. Much of the qualitative discussion in this chapter is based upon the van der Waals perspective. A complete review is given by

D. Chandler, J. D. Weeks, and H. C. Andersen, *Science* **220,** 787 (1983).

Appendix

Monte Carlo Program for Hard Disks*

```
REM***MONTE CARLO
REM***an input file "data" is needed where the first n lines contain
REM***the coordinates of the disks, the next line contains the total
REM***number of passes, and the last mg lines the total number found
REM***in each g(r) shell. "data" is updated with each run.
REM****"del" is the maximum step size. "n" is the number of disks.
REM***"rho" is the density of disks.  The screen is printed every "mcopy"
REM***times. "mg" is the number of divisions in G(r). "pas" is the number
REM*** of passes. "rgmax" is the maximum of the range G(r) is calculated
REM*** over.
del=.05
n=20
rho=.7
sig=(rho/n)^.5
rcir%=FIX(100*sig)
mcopy=1000
mg=15
pas=100
rmax=4*sig
rgmax=2.5
srgmax=rgmax*sig
dgr=(rgmax-1)/mg
ppres =rho*3.141593/2
pac = 100/n
DIM f(15),fmn(15),r(15),ddr(15),g(15)
FOR i=1 TO mg
r(i)=1+(i-.5)*dgr
NEXT i
dd=3.14*(n-1)*rho
FOR i=1 TO mg
ddr(i)=dd*((1+(i)*dgr)^2-(1+(i-1)*dgr)^2)
NEXT i
t$="monte carlo"
IN:
CALL HIDECURSOR
LET today$=DATE$
MENU 1,0,1," Monte Carlo
```

* This program was written by John D. McCoy.

```
MENU 2,0,1,""
MENU 3,0,1,"Hard Disks "
MENU 4,0,1,"        "
MENU 5,0,1,today$
WINDOW 1,t$,(0,20)-(550,350),3
DIM x(20),y(20)
OPEN "data" FOR INPUT AS #1
RANDOMIZE TIMER
FOR I=1 TO n
INPUT #1,x(I),y(I)
NEXT I
INPUT #1,pps
FOR i=1 TO mg
INPUT #1,f(i)
NEXT i
CLOSE #1
drw:
CALL MOVETO(240,25)
PRINT 5,
CALL MOVETO(240,65)
PRINT 4,
CALL MOVETO(240,105)
PRINT 3,
CALL MOVETO(240,145)
PRINT 2,
CALL MOVETO(240,185)
PRINT 1,
CALL MOVETO(240,225)
PRINT 0,
CALL MOVETO(240,125)
G$="G"
PRINT G$,
CALL MOVETO(259,240)
PRINT 1,
CALL MOVETO(339,240)
PRINT 2,
CALL MOVETO(419,240)
PRINT 3,
CALL MOVETO(359,255)
r$="r"
PRINT r$,
CALL MOVETO (20,20)
CALL LINE (0,200)
CALL LINE (200,0)
CALL LINE (0,-200)
CALL LINE (-200,0)
CALL MOVETO(270,20)
CALL LINE (0,200)
CALL LINE(160,0)
CALL MOVETO(270,20)
CALL LINE (-5,0)
CALL MOVETO(270,60)
CALL LINE (-5,0)
CALL MOVETO(270,100)
CALL LINE (-5,0)
CALL MOVETO(270,140)
CALL LINE (-5,0)
CALL MOVETO(270,180)
CALL LINE (-5,0)
CALL MOVETO(270,220)
CALL LINE (-5,0)
CALL MOVETO(270,220)
CALL LINE (0,5)
CALL MOVETO(310,220)
CALL LINE (0,5)
CALL MOVETO(350,220)
CALL LINE (0,5)
CALL MOVETO(390,220)
CALL LINE (0,5)
CALL MOVETO(430,220)
CALL LINE (0,5)
CALL MOVETO(40,250)
pas$="passes="
PRINT pas$,pps,
```

```
CALL MOVETO(40,265)
mov$="particles="
PRINT mov$,n,
CALL MOVETO(40,280)
den$="density="
PRINT den$,rho,
g1$ = "G(r=1) ="
pre$ ="PB/rho ="
ac$ = "%accept="
FOR i=1 TO n
xx=200*x(i)+20
yy=200*y(i)+20
CIRCLE (xx,yy),rcir%,33
NEXT i
FOR k=1 TO pas
acc% = 0
FOR j=1 TO n
r=1-2*RND(1)
xn=x(j)+del*r
r=1-2*RND(1)
yn=y(j)+del*r
xxo1=x(j)
yyo1=y(j)
xxo=200*xxo1+20
yyo=200*yyo1+20
FOR jj=1 TO mg
fmn(jj)=f(jj)
NEXT jj
FOR ij=1 TO n
IF ij=j THEN GOTO 10
rx= x(ij)-xn
ry= y(ij)-yn
IF rx>.5 THEN rx=rx-1 ELSE IF rx<-.5 THEN rx=rx+1
IF ry>.5 THEN ry=ry-1 ELSE IF ry<-.5 THEN ry=ry+1
r=(rx^2+ry^2 )^.5
IF r<sig THEN GOTO new1
IF r>srgmax  THEN GOTO 10
xxx=((r/sig)-1)/dgr
ii=FIX(xxx)+1
fmn(ii)=fmn(ii)+1
10 :
NEXT ij
acc% = acc% + 1
GOTO new2
NEW1:
xn =xxo1
yn=yyo1
FOR jj=1 TO mg
fmn(jj)=f(jj)
NEXT jj
FOR ij=1 TO n
IF ij=j THEN GOTO 20
rx= x(ij)-xn
ry= y(ij)-yn
IF rx>.5 THEN rx=rx-1 ELSE IF rx<-.5 THEN rx=rx+1
IF ry>.5 THEN ry=ry-1 ELSE IF ry<-.5 THEN ry=ry+1
r=(rx^2+ry^2 )^.5
IF r>srgmax  THEN GOTO 20
xxx=((r/sig)-1)/dgr
ii=FIX(xxx)+1
fmn(ii)=fmn(ii)+1
20 :
NEXT ij
NEW2
FOR jj=1 TO mg
f(jj)=fmn(jj)
NEXT jj
x(j)=xn
y(j)=yn
IF x(j)<0 THEN x(j)=x(j)+1 ELSE IF x(j)>1 THEN x(j)=x(j)-1
IF y(j)<0 THEN y(j)=y(j)+1 ELSE IF y(j)>1 THEN y(j)=y(j)-1
xx=200*x(j)+20
yy=200*y(j)+20
CIRCLE (xxo,yyo),rcir%,30
CIRCLE (xx,yy),rcir%,33
```

```
NEXT i
CALL MOVETO (20,20)
CALL LINE (0,200)
CALL LINE (200,0)
CALL LINE (0,-200)
CALL LINE (-200,0)
pps=pps+1
FOR jj=1 TO mg
g(jj)=f(jj)/(ddr(jj)*pps)
NEXT jj
CALL MOVETO((270+(r(1)-1)*80),(220-g(1)*40))
rec%(0)=0
rec%(1)=271
rec%(2)=219
rec%(3)=550
CALL ERASERECT (VARPTR(rec%(0)))
FOR jj=1 TO mg-1
xg=(r(jj+1)-r(jj))*80
yg=(g(jj)-g(jj+1))*40
CALL LINE(xg,yg)
NEXT jj
gcont = 1.5*g(1) - .5*g(2)
pres = 1 + ppres*gcont
REM**CALL MOVETO(240,280)
REM**PRINT g1$,gcont,
REM**CALL MOVETO(240,295)
REM**PRINT pre$,pres,
pacc = acc%*pac
REM**CALL MOVETO(40,295)
REM**PRINT ac$,pacc,
CALL MOVETO(40,250)
PRINT pas$,pps,
pptest=mcopy*FIX(pps/mcopy)
IF pptest=FIX(pps) THEN LCOPY
NEXT k
 out:
OPEN "data" FOR OUTPUT AS #2
FOR I=1 TO n
xn=10000*x(i)
yn=10000*y(i)
xn=FIX(xn)/10000
yn=FIX(yn)/10000
WRITE #2,xn,yn
NEXT i
WRITE #2,pps
FOR i=1 TO mg
WRITE #2,f(i)
NEXT i
CLOSE #2
WINDOW CLOSE 1
MENU RESET
END
```

The following is an example of "data".

```
1066,.6547
1701,.3779
8325,.2036
5354,.653
3416,.5222
5394,.8737
8991,.8427
4555,.3365
0682,.061
6273,.4221
7359,.0271
8751,.6159
2997,.9926
5054,.1171
1676,.8473
7247,.7575
3633,.8048
993,.3025
8265,.3979
2705,.1882
```

CHAPTER 8

Statistical Mechanics of Non-Equilibrium Systems

Up to this point, we have used statistical mechanics to describe reversible time-independent equilibrium properties. Now we move to new ground, and consider irreversible time-dependent non-equilibrium properties. An example of a non-equilibrium property is the rate at which a system absorbs energy when it is stirred by an external force. If the stirring is done with one frequency—a monochromatic disturbance—this example is the absorption spectrum of a material. Another example of a non-equilibrium property is the relaxation rate by which a system reaches equilibrium from a prepared non-equilibrium state.

Our discussion of these properties will be confined to systems close to equilibrium (in a sense made precise below). In this regime, the non-equilibrium behavior of macroscopic systems is described by *linear response theory*. This theory is the subject of this chapter.

The cornerstone of linear response theory is the *fluctuation-dissipation theorem*. This theorem is a relationship connecting relaxation and rates of absorption to the correlations between fluctuations that occur spontaneously at different times in equilibrium systems. In the development we pursue in this chapter, we first describe the theorem as a postulate and illustrate its use with two examples. In one, we analyze the rates of chemical relaxation for isomerization reactions. In the other we consider diffusive motion of particles in a fluid.

In both examples, we observe that relaxation is characterized by phenomenological parameters—a unimolecular rate constant in the first case, and a transport coefficient in the second. We will see that with the aid of the fluctuation-dissipation theorem, we can relate these phenomenological parameters to the nature of microscopic dynamics and fluctuations.

After illustrating its utility, the theorem is then derived in Sec. 8.5. The general structure of linear response theory is mapped out in Sec. 8.6 when we introduce the concept of a *response function*. With this concept, we then develop the description of dissipation and absorption experiments in Sec. 8.7. Finally, in Sec. 8.8, we describe a simple microscopic model system that exhibits the phenomena of relaxation and dissipation. This model leads to a stochastic description of dynamics known as the *Langevin equation*, and in deriving this equation we learn about *friction* and its connection to fluctuations.

This is a long list of important topics. But the discussion remains self-contained, and the mathematics employed is no more difficult than what you encountered in earlier chapters.

8.1 Systems Close to Equilibrium

To begin the discussion, we require a few elementary concepts concerning dynamics and some definitions for describing systems that are close to or slightly removed from equilibrium. By "slightly removed" we mean that the deviations from equilibrium are linearly related to the perturbations that remove the system from equilibrium. For example, consider an aqueous electrolyte solution. At equilibrium, there is no net flow of charge; the average current, $\langle j \rangle$, is zero. At some time, $t = t_1$, an electric field of strength $\mathscr{E}$ is applied, and the charged ions begin to flow. At time $t = t_2$, the field is then turned off. Let $\bar{j}(t)$ denote the observed current as a function of time. We will use the bar to denote the non-equilibrium ensemble average. The observed current is illustrated in Fig. 8.1. The non-equilibrium

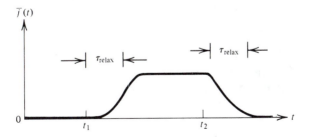

Fig. 8.1. Non-equilibrium current produced by a field applied between times t_1 and t_2.

behavior, $\bar{j}(t) \neq 0$, is linear if $\bar{j}(t)$ is proportional to $\mathcal{E}$. That is,

$$\bar{j}(t; \lambda \mathcal{E}) = \lambda \bar{j}(t; \mathcal{E}).$$

A less general way of talking about linear behavior focuses attention on thermodynamic forces or affinities rather than external fields. For example, we know that the presence of a gradient in chemical potential is associated with mass flow. Thus, for small enough gradients (i.e., small enough displacements from equilibrium) the current or mass flow should be proportional to the gradient:

$$\bar{j} \propto \nabla(\mu/T).$$

The proportionality must break down when the gradients become large. Then quadratic and perhaps higher order terms will be needed. Of course, even in the linear regime, the proportionality between $\bar{j}(t)$ and the gradient cannot be precisely correct since $\bar{j}(t)$ will probably lag in time from the behavior of the gradient. The time lag will be of the order of τ_{relax}, which is usually negligible for macroscopic non-equilibrium properties.

The meaning of a non-equilibrium ensemble average deserves some comment. Note that once initial conditions are specified, the determinism of mechanics fixes the behavior at all future times. Statistics or ensemble averages arise when we average over initial conditions according to the distribution corresponding to the observed system. To illustrate this idea, let A stand for some dynamical variable (or operator in quantum mechanics). For a classical system, A depends upon time via the time dependence of the coordinates and momenta in the system:

$$A(t) = A[r^N(t), p^N(t)].$$

But, the phase space point $[r^N(t), p^N(t)]$ is determined by integrating Newton's laws from time zero when the phase space point was $[r^N(0), p^N(0)] \equiv (r^N, p^N)$. Thus, we write

$$A(t) = A(t; \underbrace{r^N, p^N}).$$
$$ \text{initial conditions}$$

One of the basic ideas of statistical mechanics is that we do not observe $A(t)$ directly. Rather, we observe an average over all experimental possibilities of $A(t)$. The various possibilities can be samples from a distribution of initial conditions. Let $F(r^N, p^N)$ denote this distribution. Then

$$\bar{A}(t) = \int dr^N \, dp^N \, F(r^N, p^N) A(t; r^N, p^N).$$

The corresponding quantum mechanical expression is obtained by noting that the state of a system at time t, $|\psi, t\rangle$ is found uniquely in terms of the state at time zero, $|\psi\rangle$, by integrating Schrödinger's equation

$$i\hbar \frac{\partial}{\partial t} |\psi, t\rangle = \mathcal{H} |\psi, t\rangle.$$

But there is a statistical weight of initial states. Thus

$$\bar{A}(t) = \sum_{\psi} w_\psi \langle \psi, t| A |\psi, t\rangle,$$

where w_ψ is the weight for the initial state $|\psi\rangle$.

Stationary systems are those for which $\langle A(t)\rangle$ is independent of t for all possible A. Equilibrium systems are stationary systems. That is,

$$\langle A(t)\rangle = \langle A\rangle.$$

8.2 Onsager's Regression Hypothesis and Time Correlation Functions

When left undisturbed, a non-equilibrium system will relax to its terminal thermodynamic equilibrium state. Schematically, we might view the preparation and relaxation of a non-equilibrium experiment as shown in Fig. 8.2. When not far from equilibrium, the relaxation will be governed by a principle first enunciated in 1930 by Lars Onsager in his remarkable *regression hypothesis*: The relaxation of macroscopic non-equilibrium disturbances is governed by the same laws as the regression of spontaneous microscopic fluctuations in an

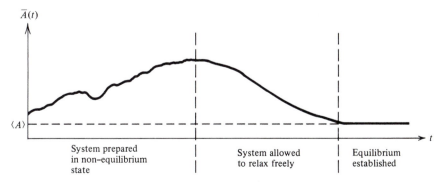

Fig. 8.2. Preparation and relaxation.

equilibrium system. This cryptic sounding principle is the cornerstone for nearly all modern work in time-dependent statistical and thermal physics. It earned Onsager the 1968 Nobel Prize in Chemistry. Present-day researchers recognize that the regression hypothesis is an important consequence of a profound theorem of mechanics: the fluctuation-dissipation theorem proved by Callen and Welton in 1951. Indeed, it is difficult to write down the precise meaning of the regression hypothesis without stating the theorem. Though Onsager never expressed his ideas explicitly in terms of the fluctuation-dissipation theorem, one suspects that he already knew the theorem and its derivation more than 20 years before others discovered the general proof.

To describe the quantitative meaning of the hypothesis, we need to talk about correlations of spontaneous fluctuations. This is done with the language of *time correlation functions*. Let $\delta A(t)$ denote the instantaneous deviation or fluctuation in $A(t)$ from its time-independent equilibrium average, $\langle A \rangle$; that is,

$$\delta A(t) = A(t) - \langle A \rangle.$$

Its time evolution is governed by microscopic laws. For classical systems,

$$\delta A(t) = \delta A(t; r^N, p^N) = \delta A[r^N(t), p^N(t)].$$

Unless A is a constant of the motion (e.g., the energy), $A(t)$ will look chaotic even in an equilibrium system. This behavior is depicted in Fig. 8.3. While the equilibrium average of $\delta A(t)$ is "uninteresting" (i.e., $\langle \delta A \rangle = 0$), one can obtain nonchaotic information by considering the equilibrium correlations between fluctuations at different times. The correlation between $\delta A(t)$ and an instantaneous or spontaneous fluctuation at time zero is

$$C(t) = \langle \delta A(0) \delta A(t) \rangle = \langle A(0)A(t) \rangle - \langle A \rangle^2.$$

Once again, the averaging is to be performed over initial conditions.

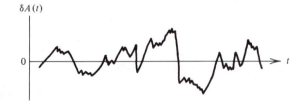

Fig. 8.3. Spontaneous fluctuations of $A(t)$ in an equilibrium system.

Thus, for a classical system

$$C(t) = \int dr^N \, dp^N f(r^N, p^N) \, \delta A(0; r^N, p^N) \, \delta A(t; r^N, p^N),$$

where $f(r^N, p^N)$ is the equilibrium phase space distribution function.

In an equilibrium system, the correlations between dynamical variables at different times should depend upon the separation between these times only and not the absolute value of time. Thus,

$$C(t) = \langle \delta A(t') \, \delta A(t'') \rangle, \quad \text{for} \quad t = t'' - t'.$$

As a special case,

$$\begin{aligned} C(t) &= \langle \delta A(0) \, \delta A(t) \rangle \\ &= \langle \delta A(-t) \, \delta A(0) \rangle. \end{aligned}$$

Now, switch the order of the two averaged quantities giving

$$\begin{aligned} C(t) &= \langle \delta A(0) \, \delta A(-t) \rangle \\ &= C(-t). \end{aligned}$$

[Note: This last manipulation assumes $A(0)$ and $A(-t)$ commute. That's all right in a classical system but not necessarily correct quantum mechanically.]

At small times,

$$C(0) = \langle \delta A(0) \, \delta A(0) \rangle = \langle (\delta A)^2 \rangle.$$

At large times, $\delta A(t)$ will become uncorrelated to $\delta A(0)$. Thus

$$C(t) \rightarrow \langle \delta A(0) \rangle \langle \delta A(t) \rangle, \quad \text{as } t \rightarrow \infty.$$

and since $\langle \delta A \rangle = 0$,

$$C(t) \rightarrow 0, \quad \text{as } t \rightarrow \infty.$$

This decay of correlations with increasing time is the "regression of spontaneous fluctuations" referred to in Onsager's hypothesis.

With the aid of the ergodic principle, there is another way of expressing the averages implied by the pointed brackets of a correlation function. In particular, imagine the behavior of A as a function of time during a long trajectory, a section of which is illustrated in Fig. 8.3. With this figure in mind, consider the correlation between the values of δA at two times t' and t'', where $t = t'' - t'$. There are an infinity of such pairs of times, and we can average over them. According to the ergodic principle, this time average will be the same as averaging over the ensemble of initial conditions for short trajectories (each of time duration t). In other

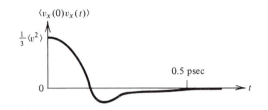

Fig. 8.4. Velocity correlation function for a liquid.

words,

$$\langle \delta A(0)\, \delta A(t)\rangle = \lim_{\tau\to\infty} \frac{1}{\tau} \int_0^\tau d\bar{t}\, \delta A(\bar{t}+t')\, \delta A(\bar{t}+t''), \qquad t = t'' - t'.$$

Here, τ represents the time period of the long trajectory. The limit $\tau \to \infty$ emphasizes that the system must be observed for a long enough time so that all of phase space can be properly sampled by the single trajectory.

As an example of what a time correlation function looks like, consider the velocity autocorrelation function for a simple atomic fluid:

$$C(t) = \langle v_x(0)v_x(t)\rangle = \tfrac{1}{3}\langle \mathbf{v}(0)\cdot \mathbf{v}(t)\rangle,$$

where $v_x(t)$ is the x-component of the velocity of a tagged particle in the fluid. At liquid densities, a qualitative picture of $C(t)$ is shown in Fig. 8.4. Another example is the orientational correlation function

$$C(t) = \langle u_z(0)u_z(t)\rangle = \tfrac{1}{3}\langle \mathbf{u}(0)\cdot \mathbf{u}(t)\rangle,$$

where $\mathbf{u}(t)$ is the unit vector along the principal axis of a tagged molecule, and $u_z(t)$ is its projection on the z-axis of a lab fixed coordinate system. For CO gas, Fig. 8.5 shows what the correlation function looks like. For the liquid phase, it looks like the function sketched in Fig. 8.6.

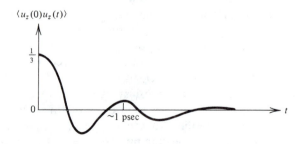

Fig. 8.5. Orientational correlation function for a gas.

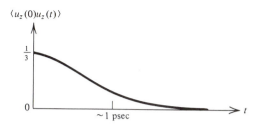

Fig. 8.6. Orientational correlation function for a liquid.

Exercise 8.1 Discuss the physical reason for the differences between the gas phase and liquid phase behavior of $\langle u_z(0)u_z(t)\rangle$.

Exercise 8.2 Draw a picture of $\langle u_z(0)u_z(t)\rangle$ for CO in a solid phase. Discuss the physical reasoning used to draw the picture.

Armed with the notation of time correlation functions, we can write down the mathematical meaning of the Onsager regression hypothesis. Imagine at time $t = 0$, a system that has been prepared in a non-equilibrium state is allowed to relax to equilibrium. Onsager's principle states that in the linear regime, the relaxation obeys

$$\frac{\Delta \bar{A}(t)}{\Delta \bar{A}(0)} = \frac{C(t)}{C(0)},$$

where

$$\Delta \bar{A}(t) = \bar{A}(t) - \langle A \rangle = \overline{\delta A(t)}$$

and

$$C(t) = \langle \delta A(0)\, \delta A(t)\rangle.$$

We will derive these equations later in Sec. 8.5. A physical motivation very nearly correct and quite close to the spirit of the mathematical derivation is the following: The correlation of $A(t)$ with $A(0)$ in an equilibrium system is the same as an average of $A(t)$ given that a certain specific fluctuation occurred at $t = 0$. Such specificity corresponds to a non-equilibrium distribution of initial phase space points. In other words, in a system close to equilibrium, we cannot distinguish between spontaneous fluctuations and deviations from equilibrium that are externally prepared. Since we cannot distinguish, the relaxation of $\langle \delta A(0)\, \delta A(t)\rangle$ should indeed coincide with the decay to equilibrium of $\Delta \bar{A}(t)$.

Exercise 8.3 Show that

$$C(t) = \langle A \rangle \Delta \bar{A}(t),$$

where the non-equilibrium distribution $F(r^N, p^N)$ in this case is

$$F(r^N, p^N) = \langle A \rangle^{-1} f(r^N, p^N) A(r^N, p^N).$$

Here, $f(r^N, p^N)$ is the equilibrium phase space distribution function—that is,

$$f(r^N, p^N) \propto \exp[-\beta \mathcal{H}(r^N, p^N)].$$

Before considering the systematic derivation of Onsager's principle, let us regard it as a postulate and turn to two illustrative applications of the principle.

8.3 Application: Chemical Kinetics

Consider the chemical reaction

$$A \rightleftarrows B,$$

where the A and B species are present in our system at very low concentrations. Let $c_A(t)$ and $c_B(t)$ denote the observed concentrations of species A and B, respectively. Reasonable phenomenological rate equations for the reaction are

$$\frac{dc_A}{dt} = -k_{BA} c_A(t) + k_{AB} c_B(t)$$

and

$$\frac{dc_B}{dt} = k_{BA} c_A(t) - k_{AB} c_B(t),$$

where k_{BA} and k_{AB} are the forward and back rate constants, respectively. Note that $c_A(t) + c_B(t)$ is a constant in this model. Also note that the equilibrium concentrations, $\langle c_A \rangle$ and $\langle c_B \rangle$, must obey the *detailed balance condition*: $0 = -k_{BA} \langle c_A \rangle + k_{AB} \langle c_B \rangle$. That is,

$$K_{eq} \equiv (\langle c_B \rangle / \langle c_A \rangle) = (k_{BA}/k_{AB}).$$

The solutions to the rate equations yield

$$\Delta c_A(t) = c_A(t) - \langle c_A \rangle = \Delta c_A(0) \exp(-t/\tau_{rxn}),$$

where

$$\tau_{rxn}^{-1} = k_{AB} + k_{BA}.$$

Exercise 8.4 Verify this result.

Suppose that n_A is the dynamical variable for which

$$\overline{n_A}(t) \propto c_A(t).$$

Then, according to the fluctuation-dissipation theorem or regression hypothesis

$$\Delta c_A(t)/\Delta c_A(0) = \langle \delta n_A(0)\, \delta n_A(t)\rangle / \langle (\delta n_A)^2\rangle.$$

As a result,

$$\exp\left(-t/\tau_{rxn}\right) = \langle \delta n_A(0)\, \delta n_A(t)\rangle / \langle (\delta n_A)^2\rangle.$$

This is indeed a remarkable result. The left-hand side contains the phenomenological rate constant—that is, τ_{rxn}^{-1}—and the right-hand side is completely defined in terms of microscopic mechanics and ensemble averages. Thus, the regression hypothesis provides a method by which one may compute a rate constant from microscopic laws.

Of course, there are two difficulties. First, we must determine the dynamical variable n_A. Second, the exact integration of equations of motion and averaging to obtain $\langle \delta n_A(0)\, \delta n_A(t)\rangle$ are formidable tasks for all but the simplest systems.

To identify n_A, we must provide a microscopic definition of the molecular species A. This task is particularly simple if the reaction can be described by a single reaction coordinate, q.

A schematic picture of the potential energy "surface" for the reaction is given in Fig. 8.7. The activated state, located at $q = q^*$, provides a convenient dividing surface separating A and B species. That is, $q < q^*$ corresponds to species A, and $q > q^*$ corresponds to species B. (In the most general case, one must adopt a rule that says a certain species or composition of species corresponds to a region in phase space. The particular rule adopted must depend upon the experimental criteria used to distinguish chemical species.) Thus, we

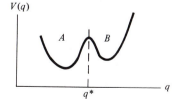

Fig. 8.7. Potential energy surface for the reaction coordinate.

let

$$n_A(t) = H_A[q(t)],$$

where

$$H_A[z] = 1, \quad z < q^*$$
$$= 0, \quad z > q^*.$$

Note

$$\langle H_A \rangle = x_A = \langle c_A \rangle / (\langle c_A \rangle + \langle c_B \rangle)$$

and

$$\langle H_A^2 \rangle = \langle H_A \rangle = x_A.$$

Hence

$$\langle (\delta H_A)^2 \rangle = x_A(1 - x_A)$$
$$\equiv x_A x_B.$$

Exercise 8.5 Verify these results.

According to the fluctuation-dissipation theorem, we now have

$$\exp(-t/\tau_{rxn}) = (x_A x_B)^{-1}[\langle H_A(0)H_A(t) \rangle - x_A^2].$$

To analyze the consequences of this relationship, we take a time derivative to obtain

$$\tau_{rxn}^{-1} \exp(-t/\tau_{rxn}) = -(x_A x_B)^{-1}\langle H_A(0)\dot{H}_A(t) \rangle,$$

where the dot denotes a time derivative. Since $\langle A(t)A(t') \rangle = \langle A(0)A(t' - t) \rangle = \langle A(t - t')A(0) \rangle$, we have

$$-\langle H_A(0)\dot{H}_A(t) \rangle = \langle \dot{H}_A(0)H_A(t) \rangle.$$

Exercise 8.6 Derive this result.

Furthermore,

$$\dot{H}_A[q] = \dot{q}\frac{d}{dq}H_A[q] = -\dot{q}\delta(q - q^*).$$

Hence

$$-\langle H_A(0)\dot{H}_A(t) \rangle = -\langle \dot{q}(0)\,\delta[q(0) - q^*]H_A[q(t)] \rangle$$
$$= \langle \dot{q}(0)\,\delta[q(0) - q^*]H_B[q(t)] \rangle,$$

where the second equality is obtained from

$$H_B[z] = 1 - H_A[z] = 1, \quad z > q^*,$$
$$= 0, \quad z < q^*,$$

and the fact that

$$\langle \dot{q}(0)\delta[q(0) - q^*]\rangle = 0.$$

This last result is true because the velocity is an odd vector function and the equilibrium ensemble distribution of velocities is even and uncorrelated with configurations. Combining these equations gives

$$\tau_{rxn}^{-1} \exp\left(-t/\tau_{rxn}\right) = (x_A x_B)^{-1}\langle v(0)\delta[q(0) - q^*]H_B[q(t)]\rangle,$$

where $v(0) = \dot{q}(0)$.

But this equality cannot be correct for all times. The left-hand side is a simple exponential. The right-hand side is an average flux crossing the "surface" at $q = q^*$ given that the trajectory ends up in state B. For short times, we expect transient behavior that should not correspond to the exponential macroscopic decay. This does not mean that the regression hypothesis is wrong. Rather, the phenomenological rate laws we have adopted can only be right after coarse graining in time. In other words, the phenomenology can only be right on a time scale that does not resolve the short time transient relaxation. On that time scale, let Δt be a small time. That is,

$$\Delta t \ll \tau_{rxn},$$

but at the same time

$$\Delta t \gg \tau_{mol},$$

where τ_{mol} is the time for transient behavior to relax. For such times, $\exp\left(-\Delta t/\tau_{rxn}\right) \approx 1$, and we obtain

$$\tau_{rxn}^{-1} = (x_A x_B)^{-1}\langle v(0)\delta[q(0) - q^*]H_B[q(\Delta t)]\rangle$$

or

$$k_{BA} = x_A^{-1}\langle v(0)\delta[q(0) - q^*]H_B[q(\Delta t)]\rangle.$$

To illustrate the transient behavior we have just described, let

$$k_{BA}(t) = x_A^{-1}\langle v(0)\delta[q(0) - q^*]H_B[q(t)]\rangle.$$

Exercise 8.7 Show that

$$k_{BA}(0) = (1/2x_A)\langle |v|\rangle \langle \delta(q - q^*)\rangle$$

and verify that this initial rate is precisely that of the transition state theory approximation,

$$k_{BA}^{(TST)} = (1/x_A)\langle v(0)\delta[q(0) - q^*]H_B^{(TST)}[q(t)]\rangle,$$

where

$$H_B^{(TST)}[q(t)] = 1, \qquad v(0) > 0,$$
$$= 0, \qquad v(0) < 0.$$

In other words, *transition state theory* assumes that all trajectories initially heading from the transition state toward region B will indeed end up in region B for a long time. Similarly, those with $q(0) = q^*$ and $\dot{q}(0) < 0$ will not go to region B. This assumption would be correct if no trajectories recrossed the transition state after a short time.

Then $k_{BA}(t)$ appears as shown in Fig. 8.8.

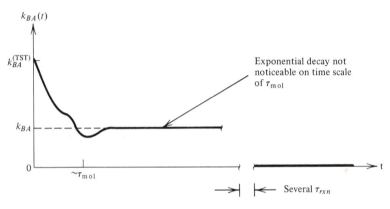

Fig. 8.8. The reactive flux correlation function.

When taken literally for all times, the phenomenological model acts as if the reaction occurs instantaneously. But, once started, a reaction does take place over a finite time period, and observations that monitor times as short as those in which the reactions occur can observe deviations from phenomenology, in particular the transient relaxation depicted in Fig. 8.8.

In closing this discussion, we note that if on any time scale, $k_{BA}(t)$ does not exhibit a plateau, then the rate law in which k_{BA} is a constant is not a valid description of the chemical kinetics.

Exercise 8.8* Show that $k_{BA}^{(TST)} \geq k_{BA}(t)$. [*Hint:* Consider the contribution to $k_{BA}(t)$ from trajectories that recross the surface $q = q^*$. See Exercise 8.7.]

8.4 Another Application: Self-Diffusion

Consider a solute that is present in very low concentrations in a fluid solvent. (The solute can even be a few tagged solvent molecules.) We let $n(\mathbf{r}, t)$ denote the non-equilibrium density of solutes at position $\mathbf{r}$ at time t. That is,

$$n(\mathbf{r}, t) = \bar{\rho}(\mathbf{r}, t),$$

where $\rho(\mathbf{r}, t)$ is the instantaneous density at position $\mathbf{r}$.

Exercise 8.9* Show that

$$\rho(\mathbf{r}, t) = \sum_{j=1}^{N} \delta[\mathbf{r} - \mathbf{r}_j(t)],$$

where $\mathbf{r}_j(t)$ is the position of the jth solute molecule at time t, and $\delta[\mathbf{r} - \mathbf{r}_j(t)]$ is the three-dimensional Dirac delta function which is zero except when $\mathbf{r} = \mathbf{r}_j(t)$.

The density in a region of the fluid changes in time because particles flow in and out of the region. However, neglecting chemical reactions, molecules cannot be created or destroyed. In other words, the total number of solute molecules in the total system is a constant. This conservation leads to an *equation of continuity*:

$$\frac{\partial}{\partial t} n(\mathbf{r}, t) = -\nabla \cdot \mathbf{j}(\mathbf{r}, t),$$

where $\mathbf{j}(\mathbf{r}, t)$ is the non-equilibrium average flux of solute particles at position $\mathbf{r}$ at time t.

To derive the equation of continuity, consider an arbitrary volume Ω within the system. See Fig. 8.9. The total number of solute particles inside the volume is

$$\bar{N}_\Omega(t) = \int_\Omega d\mathbf{r}\, n(\mathbf{r}, t).$$

Since mass flow is the only mechanism for changing $N_\Omega(t)$, we have

$$d\bar{N}_\Omega/dt = -\int_S d\mathbf{S} \cdot \mathbf{j}(\mathbf{r}, t) = -\int_\Omega d\mathbf{r}\, \nabla \cdot \mathbf{j}(\mathbf{r}, t),$$

where the first integral is over the surface S that surrounds the

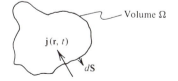

Fig. 8.9. Flux into a volume.

volume, and the minus sign appears because $\mathbf{j} \cdot d\mathbf{S}$ is positive when flow is *out* of the system. Of course, we also have

$$d\bar{N}_\Omega / dt = \int_\Omega d\mathbf{r}\, \partial n(\mathbf{r}, t) / \partial t.$$

Thus,

$$0 = \int_\Omega d\mathbf{r}[\partial n(\mathbf{r}, t) / \partial t + \nabla \cdot \mathbf{j}(\mathbf{r}, t)].$$

This equation must hold for any volume Ω. As a result, the integrand must be zero for all $\mathbf{r}$. By setting the integrand to zero, we obtain the equation of continuity.

Exercise 8.10* Identify the dynamical variable whose non-equilibrium average is $\mathbf{j}(\mathbf{r}, t)$. Show that the equation of continuity can be obtained directly from the delta-function of $\rho(\mathbf{r}, t)$. Discuss the relationship between the equation of continuity and the equation $\dot{H}_A[q] = -\dot{q}\delta(q - q^*)$ encountered in the examination of an activated process.

The macroscopic thermodynamic mechanism for mass flow is a chemical potential gradient or equivalently, for a dilute solution, a solute concentration gradient. Hence, a reasonable phenomenological relationship is

$$\mathbf{j}(\mathbf{r}, t) = -D\nabla n(\mathbf{r}, t),$$

where D is called the self-*diffusion constant*. This relationship is called *Fick's law*. When combined with the equation of continuity, it yields

$$\partial n(\mathbf{r}, t) / \partial t = D\nabla^2 n(\mathbf{r}, t),$$

which is also called Fick's law.

Solutions to this partial differential equation describe how concentration gradients relax in terms of a parameter, the diffusion constant. To learn how this transport coefficient is related to microscopic dynamics, we now consider the correlation function

$$C(\mathbf{r}, t) = \langle \delta\rho(\mathbf{r}, t)\delta\rho(\mathbf{0}, 0) \rangle.$$

According to Onsager's regression hypothesis, $C(\mathbf{r}, t)$ obeys the same equation as $n(\mathbf{r}, t)$. That is

$$\partial C(\mathbf{r}, t) / \partial t = D\nabla^2 C(\mathbf{r}, t).$$

But note that $\langle \rho(\mathbf{r}, t)\rho(\mathbf{0}, 0)\rangle$ is proportional to

$P(\mathbf{r}, t) =$ conditional probability distribution that a solute particle is at $\mathbf{r}$ at time t given that the particle was at the origin at time zero.

Exercise 8.11* Explain this proportionality.

As a result,

$$\partial P(\mathbf{r}, t)/\partial t = D\nabla^2 P(\mathbf{r}, t).$$

Since $P(\mathbf{r}, t)$ or $C(\mathbf{r}, t)$ is well defined at the molecular level, these differential equations provide the necessary connection between the self-diffusion constant, D, and microscopic dynamics. Of course, it is impossible for equations like $\partial C/\partial t = D\nabla^2 C$ to be correct for all time or spatial variations. The error in the equation is due to the inadequacies of Fick's law.

To proceed with the analysis, we consider

$$\Delta R^2(t) = \langle |\mathbf{r}_1(t) - \mathbf{r}_1(0)|^2\rangle,$$

which is the mean squared displacement of a tagged solute molecule in a time t. Clearly,

$$\Delta R^2(t) = \int d\mathbf{r}\, r^2 P(\mathbf{r}, t).$$

Thus

$$\frac{d}{dt}\Delta R^2(t) = \int d\mathbf{r}\, r^2 \partial P(\mathbf{r}, t)/\partial t$$

$$= \int d\mathbf{r}\, r^2 D\nabla^2 P(\mathbf{r}, t)$$

$$= 6\int d\mathbf{r}\, DP(\mathbf{r}, t),$$

where the last equality is obtained by integrating by parts twice.

Exercise 8.12 Verify this result. [*Hint:* Note that with Cartesian coordinates, $r^2 = x^2 + y^2 + z^2$, and neglect the boundary terms (why is that possible?).]

Since $P(\mathbf{r}, t)$ is normalized for all times, we have

$$\frac{d}{dt}\Delta R^2(t) = 6D$$

or

$$\Delta R^2(t) = 6Dt.$$

This formula (first derived by Einstein) clarifies the physical meaning of the diffusion constant. However, it is valid only after an initial transient time [analogous to the time it takes $k_{BA}(t)$ to relax to its plateau value].

Notice the difference between diffusional motion and inertial motion. In the former, $\Delta R^2(t) \propto t$, while for inertial behavior $\Delta R(t) = (\text{velocity}) \cdot (t)$ or

$$[\Delta R(t)]_{\text{inertial}} \propto t.$$

In other words, for large times, when a particle moves inertially, it progresses much farther per unit time than it does when moving diffusively. The reason is that diffusive motion is much like a random walk where the particle is buffeted about by random fluctuating forces due to its environment. (This fact will perhaps become clearer in Sec. 8.8 where we describe a model for a particle moving under the influence of such fluctuating forces.) In the inertial regime, however, a particle does not encounter its neighbors, and it moves freely undeterred from its initial direction of motion. It takes a finite amount of time for a force to alter the initial velocity of a particle (i.e., the time derivative of the velocity is proportional to the force, "$F = ma$"). Hence, for short enough times, the correlated motion of a particle in a fluid will be inertial. The time for the inertial regime to relax is the time it takes for the diffusional regime to take over. The behavior we have just described is depicted in Fig. 8.10.

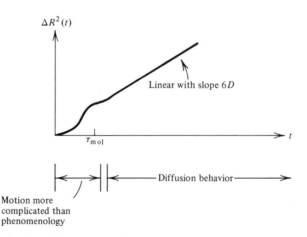

Fig. 8.10. Mean square displacement of a diffusing particle.

To finish the analysis, note that

$$\mathbf{r}_1(t) - \mathbf{r}_1(0) = \int_0^t dt' \, \mathbf{v}(t'),$$

where $\mathbf{v}(t)$ is the velocity of the tagged solute—that is, $d\mathbf{r}_1/dt = \mathbf{v}$. Thus,

$$\Delta R^2(t) = \int_0^t dt' \int_0^t dt'' \langle \mathbf{v}(t') \cdot \mathbf{v}(t'') \rangle.$$

As a result,

$$\frac{d}{dt} \Delta R^2(t) = 2 \langle \mathbf{v}(t) \cdot [\mathbf{r}_1(t) - \mathbf{r}_1(0)] \rangle$$

$$= 2 \langle \mathbf{v}(0) \cdot [\mathbf{r}_1(0) - \mathbf{r}_1(-t)] \rangle$$

$$= 2 \int_{-t}^0 \langle \mathbf{v}(0) \cdot \mathbf{v}(t') \rangle \, dt',$$

where in the second equality, we changed the origin of time. [Recall, for a stationary system, $\langle A(t)B(t') \rangle = \langle A(t+t'')B(t'+t'') \rangle$.] Further, since $\langle A(t)A(t') \rangle = C(t-t') = C(t'-t)$, the last equality can also be written as

$$\frac{d}{dt} \Delta R^2(t) = 2 \int_0^t dt' \langle \mathbf{v}(0) \cdot \mathbf{v}(t') \rangle.$$

Since the left-hand side tends to $6D$ in the limit of large t, we have the identification

$$D = (1/3) \int_0^\infty dt \, \langle \mathbf{v}(0) \cdot \mathbf{v}(t) \rangle.$$

This equation relates a transport coefficient to an integral of an autocorrelation function. Such relationships are known as *Green-Kubo formulas*.

One final remark focuses on the velocity relaxation time, τ_{relax}, which we can define as

$$\tau_{\text{relax}} = \int_0^\infty dt \, \langle \mathbf{v}(0) \cdot \mathbf{v}(t) \rangle / \langle v^2 \rangle.$$

Clearly,

$$\tau_{\text{relax}} = \beta m D.$$

It has the physical meaning of the time it takes for the diffusional regime to dominate the motion, and it is the time for non-equilibrium velocity distributions to relax to equilibrium. In other words, $\beta m D$ is the time identified as τ_{mol} in Fig. 8.10.

Exercise 8.13 A simple approximation for the velocity correlation function is to say it relaxes exponentially,

$$\langle \mathbf{v}(0) \cdot \mathbf{v}(t) \rangle \approx \langle v^2 \rangle e^{-t/\tau}.$$

Use this approximation to compute $\Delta R^2(t)$. [*Hint*: It will be helpful to first calculate $d\Delta R^2(t)/dt$.] Draw a graph of your result and show that $\Delta R^2(t)$ changes from nondiffusive to diffusive behavior after a time of the order of τ. Finally, it is an experimental fact that the self-diffusion constant for most liquids is about $10^{-5} \, \text{cm}^2/\text{sec}$. Use this value to estimate the size of τ. (You will need to specify a temperature and a typical mass of a molecule.)

8.5 Fluctuation-Dissipation Theorem

Let us now see how it is possible to derive Onsager's regression hypothesis from the principles of statistical mechanics. We will confine our attention to classical systems, though a similar treatment for the quantum mechanical case is not difficult. The analysis we carry out below is a simplified version of *linear response theory*—the theory of systems weakly displaced from equilibrium. Our principal result, a correlation function expression for $\bar{A}(t)$, is a version of the *fluctuation-dissipation theorem*.

The Hamiltonian for the system when undisturbed is $\mathcal{H}$, and the equilibrium average of a dynamical variable $A(r^N, p^N)$ would be

$$\langle A \rangle = \int dr^N \, dp^N \, e^{-\beta \mathcal{H}(r^N, p^N)} A(r^N, p^N) \Big/ \int dr^N \, dp^N \, e^{-\beta \mathcal{H}}$$

$$= \text{Tr} \, e^{-\beta \mathcal{H}} A / \text{Tr} \, e^{-\beta \mathcal{H}},$$

where the second equality introduces the "classical trace," which is simply an abbreviation for the integral over phase space variables r^N p^N. At time $t = 0$, the system is not at equilibrium, however, and we would like to see, if left undisturbed and allowed to relax to equilibrium, how the non-equilibrium $\bar{A}(t)$ relaxes to $\langle A \rangle$. It is very difficult to characterize this relaxation in a general way. However, if the deviation from equilibrium is not large, the analysis is straightforward.

To begin, it is convenient to imagine that the system was prepared in its non-equilibrium state by the application of a perturbation

$$\Delta \mathcal{H} = -fA,$$

where f is an applied field that couples to A. [For example, f might be an electric field that couples to the instantaneous dipole moments of the molecules in the system, in which case A is the total dipole, or polarization, of the system.] We will consider the case where this perturbation was applied in the distant past, and it was maintained until $t = 0$ when it was then shut off. Accordingly, the system has been prepared in such a way that the initial, $t = 0$, non-equilibrium distribution in phase space is the equilibrium distribution when the field is on. That is,

$$F(r^N p^N) \propto e^{-\beta(\mathcal{H}+\Delta\mathcal{H})}.$$

As a result, the initial value of $\bar{A}(t)$ is

$$\bar{A}(0) = \mathrm{Tr}\, e^{-\beta(\mathcal{H}+\Delta\mathcal{H})} A / \mathrm{Tr}\, e^{-\beta(\mathcal{H}+\Delta\mathcal{H})}.$$

As time evolves, $\bar{A}(t)$ changes in accord with the formula $\bar{A}(t) = \mathrm{Tr}\, FA(t)$; that is,

$$\bar{A}(t) = \mathrm{Tr}\, e^{-\beta(\mathcal{H}+\Delta\mathcal{H})} A(t) / \mathrm{Tr}\, e^{-\beta(\mathcal{H}+\Delta\mathcal{H})},$$

where $A(t) = A(t; r^N, p^N)$ is the value of A at time t given the initial phase space point $r^N p^N$, and the Hamiltonian governing the dynamics that changes $A(0)$ to $A(t)$ is $\mathcal{H}$ (not $\mathcal{H} + \Delta\mathcal{H}$ since $\Delta\mathcal{H}$ was turned off at $t = 0$).

Since the deviations of $A(t)$ from $\langle A \rangle$ are due to $\Delta\mathcal{H}$, and since we stipulated at the beginning that we will assume these deviations are small, we shall expand $\bar{A}(t)$ in a series ordered by $\Delta\mathcal{H}$. This expansion is

$$\bar{A}(t) = \mathrm{Tr}\, [e^{-\beta\mathcal{H}}(1 - \beta\Delta\mathcal{H} + \cdots)$$
$$\times A(t)] / \mathrm{Tr}\, [e^{-\beta\mathcal{H}}(1 - \beta\Delta\mathcal{H} + \cdots)].$$

{It is at this stage where the quantum mechanical analysis would be somewhat different since $\mathcal{H}$ and $\Delta\mathcal{H}$ would then be operators that, in general, would not commute, so that the expansion of $\exp[-\beta(\mathcal{H} + \Delta\mathcal{H})]$ becomes a little tricky. It's left to the reader as an exercise.} By multiplying out the terms in the expansion and collecting through linear order in $\Delta\mathcal{H}$, we find

$$\bar{A}(t) = \mathrm{Tr}\, \{e^{-\beta\mathcal{H}}[A(t) - (\beta\Delta\mathcal{H})A(t)$$
$$+ A(t)\, \mathrm{Tr}\, e^{-\beta\mathcal{H}}(\beta\Delta\mathcal{H})/\mathrm{Tr}\, e^{-\beta\mathcal{H}}]\}/\mathrm{Tr}\, e^{-\beta\mathcal{H}}$$
$$+ O((\beta\Delta\mathcal{H})^2)$$
$$= \langle A \rangle - \beta[\langle \Delta\mathcal{H}A(t)\rangle - \langle A \rangle\langle \Delta\mathcal{H}\rangle]$$
$$+ O((\beta\Delta\mathcal{H})^2),$$

where the second equality is obtained noting that due to the stationarity of an equilibrium ensemble, the average of a variable at any one time is independent of time (i.e., $\langle A(t) \rangle = \langle A \rangle$). Notice that in deriving this expansion, nothing specific was assumed about how the function $A(t) = A(t; r^N, p^N)$ evolves in time from its value evaluated at the initial phase space point r^N, p^N. We do assume, however, that its average value (i.e., its observed value) as a function of time is a well-behaved function of the initial distribution of phase space points provided that distribution is close to the equilibrium one.

Finally, we end the analysis by inserting the expression for $\Delta \mathcal{H}$. This gives

$$\Delta \bar{A}(t) = \beta f \langle \delta A(0) \delta A(t) \rangle + O(f^2),$$

where $\Delta \bar{A}(t) = \bar{A}(t) - \langle A \rangle$, and $\delta A(t)$ is the spontaneous and instantaneous fluctuation at time t, $A(t) - \langle A \rangle$. Note that since due to linear order,

$$\Delta \bar{A}(0) = \beta f \langle (\delta A)^2 \rangle,$$

the boxed equation is precisely the statement given as the regression hypothesis in Sec. 8.2.

We could also consider a $\Delta \mathcal{H}$ of the general form

$$\Delta \mathcal{H} = - \sum_i f_i A_i.$$

An important example is an external spatially dependent potential field, $\Phi_{\text{ext}}(\mathbf{r})$, which couples to the particle density at $\mathbf{r}$, $\rho(\mathbf{r})$; that is,

$$\Delta \mathcal{H} = \int d\mathbf{r} \, \Phi_{\text{ext}}(\mathbf{r}) \rho(\mathbf{r}).$$

In this example, the subscript i labels points in space; that is, we make the associations $f_i \leftrightarrow -\Phi_{\text{ext}}(\mathbf{r})$ and $A_i \leftrightarrow \rho(\mathbf{r})$. The analysis of this case gives

$$\Delta \bar{A}_j(t) = \beta \sum_i f_i \langle \delta A_i(0) \delta A_j(t) \rangle + O(f^2).$$

Exercise 8.14 Derive this result.

Exercise 8.15 Use this result to justify the application of Fick's law to analyze the density–density correlation function.

In closing this section, it is worth noting that some scientists reserve the terminology "fluctuation-dissipation theorem" for a relationship that is equivalent to but somewhat different in appearance from the one we have boxed in above. Indeed, at this stage we have demonstrated only that the relaxation of $\Delta \bar{A}(t)$ is connected to the correlations between spontaneous fluctuations occurring at different times. It is yet to be established that this relaxation is related to the rate at which energy is dissipated in a macroscopic system. To carry out a demonstration requires a bit more analysis, which we present in the next two sections.

8.6 Response Functions

The fluctuation-dissipation theorem connects the relaxation from a prepared non-equilibrium state with the spontaneous microscopic dynamics in an equilibrium system. As the derivation reveals, the theorem is valid in the linear regime—that is, for displacements not far from equilibrium. Within this regime, the result of the fluctuation-dissipation theorem can be applied much more generally than might be apparent from the specific derivation. This point is understood by introducing the concept of a *response function*.

Imagine a time-dependent perturbation in which the externally applied disturbance, $f(t)$, couples to a dynamical variable, A. In the linear regime $\Delta \bar{A}(t; f)$ satisfies

$$\Delta \bar{A}(t; \lambda f) = \lambda \Delta \bar{A}(t; f),$$

and the most general form consistent with this condition is

$$\Delta \bar{A}(t) = \int_{-\infty}^{\infty} dt' \, \chi(t, t') f(t') + O(f^2).$$

Here, $\chi(t, t')$ is the response function. It is also sometimes called the generalized susceptibility. It is a property of the equilibrium system in the absence of $f(t)$. Perhaps this fact is best understood by viewing the connection between $\Delta \bar{A}(t)$ and $f(t')$ as the first term in a Taylor series. In particular, $\Delta \bar{A}(t)$ depends upon the function $f(t)$. That is, $\Delta \bar{A}(t)$ is said to be a *functional* of $f(t)$. One may think of $f(t)$ at each point in time as another variable. When all the $f(t)$'s are zero, $\Delta \bar{A}(t) = 0$ since the system is then at equilibrium. Hence, the first nonvanishing term in a Taylor series for $\Delta \bar{A}(t)$ ordered in powers of $f(t)$ is

$$\sum_{i} [\partial \Delta \bar{A}(t) / \partial f(t_i)]_0 f(t_i)$$

where the subscript zero indicates that the derivative is evaluated with $f(t) = 0$, and the sum is over points in time, t_i. Of course, these points form a continuum, and the sum is properly an integral. In this continuum limit, the partial derivative becomes what is known as a *functional* or (variational) derivative denoted by $\delta \Delta \bar{A}(t)/\delta f(t')$. It is the rate of change of $\Delta \bar{A}(t)$ due to a change in $f(t')$ at the point in time t'. Hence, $\chi(t, t')$ is the rate of change of $\Delta \bar{A}(t)$ when the system is perturbed by an external disturbance at time t'.

Another way of viewing this idea arises from considering an impulsive disturbance. That is, $f(t)$ is applied at only one time, $t = t_0$:

$$f(t) = f_0 \delta(t - t_0).$$

Inserting this form into $\Delta \bar{A}(t; f)$ yields

$$\Delta \bar{A}(t) = f_0 \chi(t, t_0) + O(f_0^2).$$

The response function is therefore $\Delta \bar{A}(t)$ when the perturbation is an impulse. All other responses can be interpreted as linear combinations of this special case.

Two properties of $\chi(t, t')$ follow from physical considerations. First, since $\chi(t, t')$ is a function of the unperturbed equilibrium system,

$$\chi(t, t') = \chi(t - t').$$

That is, the response depends only on the time displacement from when the disturbance was applied and not on the absolute time. Second, the system cannot respond until after the disturbance is applied. That is,

$$\chi(t - t') = 0, \qquad t - t' \leq 0.$$

This property is known as *causality*. Figure 8.11 depicts schematically the behavior we have been describing.

Let us now express $\chi(t - t')$ in terms of the intrinsic dynamics of the system. $\chi(t - t')$ is independent of $f(t)$, and we are free to pick the form of $f(t)$ to make the analysis convenient. Our choice is

$$f(t) = f, \qquad t < 0$$
$$= 0, \qquad t \geq 0.$$

That is, the system is prepared at equilibrium with the Hamiltonian $\mathcal{H} - fA$, and then at time zero, $f(t)$ is turned off, and the system relaxes to the equilibrium state with Hamiltonian $\mathcal{H}$. This is precisely the experiment considered in the previous section. There, we discovered

$$\Delta \bar{A}(t) = \beta f \langle \delta A(0) \delta A(t) \rangle + O(f^2).$$

Now we consider what the result is in terms of $\chi(t - t')$.

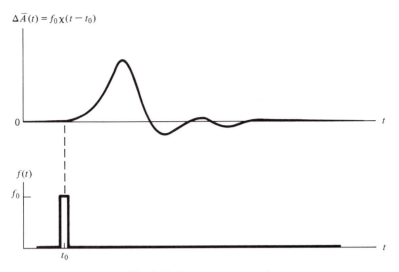

Fig. 8.11. Response to a pulse.

Since $f(t')$ is zero for $t' > 0$, we have

$$\Delta \bar{A}(t) = f \int_{-\infty}^{0} dt' \, \chi(t - t').$$

Change the variable of integration to $t - t'$,

$$\Delta \bar{A}(t) = f \int_{t}^{\infty} dt' \, \chi(t').$$

By comparing with the result of the fluctuation-dissipation theorem, and noting $\chi(t - t')$ is zero for $t < t'$, we have

$$\chi(t) = -\beta \frac{d}{dt} \langle \delta A(0) \delta A(t) \rangle, \qquad t > 0$$
$$= 0, \qquad t < 0.$$

This formula relates the general behavior of a non-equilibrium system in the linear regime to the correlations between spontaneous fluctuations at different times as they occur in the equilibrium system.

Exercise 8.16 Consider a pulse experiment where

$$\begin{aligned} f(t) &= 0, & t < t_1 \\ &= f, & t_1 < t < t_2 \\ &= 0, & t > t_2, \end{aligned}$$

and assume

$$\langle \delta A(0)\delta A(t)\rangle = \langle (\delta A)^2\rangle \exp{(-t/\tau)}.$$

Calculate and draw a sketch of $\Delta \bar{A}(t)$ as a function of time. Consider separately the cases where $\tau \ll t_2 - t_1$, and $\tau \gg t_2 - t_1$. Also calculate the energy absorbed; that is, calculate

$$-\int_{-\infty}^{\infty} dt\, \dot{f}(t)\bar{A}(t).$$

8.7 Absorption

One important class of experiments probes the absorption of energy when a system is disturbed by a monochromatic disturbance. An obvious example is standard spectroscopy (e.g., IR absorption) where $f(t)$ corresponds to an oscillating electric field and the dynamical variable A is the total dipole or polarization of the material under examination.

The time-dependent perturbation to the energy is $-f(t)A$, and its rate of change is $-(df/dt)A$. Therefore, the total energy absorbed by the system per unit time over the course of an observation

$$\text{abs} = -(1/T)\int_0^T dt\, \dot{f}(t)\bar{A}(t),$$

where T is the time duration of the observation. Alternatively, we arrive at this result by noting that if the contribution to $\mathcal{H}$ from an external field is $-fA$, then the differential work done on the system as a result of changing $\bar{A}$ is $f\, d\bar{A}$. Per unit time, this gives $f\, d\bar{A}/dt$. Hence,

$$\text{abs} = (1/T)\int_0^T dt\, f(t)(d\bar{A}/dt)$$

$$= -(1/T)\int_0^T dt\, \dot{f}(t)\bar{A}(t),$$

where in the second equality we have integrated by parts and neglected the boundary term (which is negligible for large enough T).

For a monochromatic disturbance of frequency ω, we have

$$f(t) = \text{Re}\, f_\omega e^{-i\omega t}.$$

Therefore,

$$\text{abs} = (1/T) \int_0^T dt [i\omega(f_\omega e^{-i\omega t} - f_\omega^* e^{i\omega t})/2] \bar{A}(t)$$

$$= (1/T) \int_0^T dt [i\omega(f_\omega e^{-i\omega t} - f_\omega^* e^{i\omega t})/2]$$

$$\times \left\{ \langle A \rangle + \int_{-\infty}^\infty dt' \, \chi(t') f(t - t') + O(f^2) \right\},$$

where in the second equality we have used the linear response formula for $A(t)$ and noted

$$\int_{-\infty}^\infty dt' \, \chi(t - t') f(t') = \int_{-\infty}^\infty dt' \, \chi(t') f(t - t').$$

Next, let us consider observation times that are much larger than a period of oscillation for $f(t)$; that is, $T \gg 2\pi/\omega$. For such large times,

$$(1/T) \int_0^T dt \, e^{in\omega t} = 1, \qquad n = 0,$$

$$= 0, \qquad n \neq 0.$$

Exercise 8.17 Verify that this formula is indeed correct for $\omega T \to \infty$.

By using this result in the formula for $\text{abs}(\omega)$ after expressing $f(t - t')$ in terms of $\exp[\pm i\omega(t - t')]$ and multiplying out all the terms in the integrals, we find

$$\text{abs}(\omega) = (\omega |f_\omega|^2/2) \int_{-\infty}^\infty dt \, \chi(t) \sin(\omega t),$$

where we have neglected terms higher order in f than f^2.

Exercise 8.18 Verify this result.

This equation shows that the absorption is proportional to the sine-Fourier transform of the response function. The fluctuation-dissipation theorem relates $\chi(t)$ to the derivative of $C(t) = \langle \delta A(0) \delta A(t) \rangle$. This connection lets us write

$$\int_{-\infty}^\infty dt \, \chi(t) \sin(\omega t) = -\beta \int_0^\infty dt \, [dC(t)/dt] \sin(\omega t)$$

$$= \beta\omega \int_0^\infty dt \, C(t) \cos(\omega t),$$

where we have integrated by parts. Therefore,

$$\text{abs}\,(\omega) = (\beta \omega^2 \,|f_\omega|^2/4) \int_0^\infty dt \,\langle \delta A(0)\delta A(t)\rangle \cos(\omega t).$$

Hence, the absorption spectrum is proportional to the Fourier transform of the autocorrelation function for $\delta A(t)$ in the equilibrium system. This result provides an important interpretation of spectroscopy since it demonstrates how the spectrum is related to the spontaneous dynamics in the system.

As an example, suppose $A(t)$ obeyed simple harmonic oscillator dynamics. That is,

$$d^2 A(t)/dt^2 = -\omega_0^2 A(t),$$

where ω_0 is the oscillator (angular) frequency. The solution of this linear equation in terms of the initial conditions $A(0)$ and $\dot{A}(0)$ is easy to find, and the result leads to

$$\langle \delta A(0)\delta A(t)\rangle = \langle(\delta A)^2\rangle \cos(\omega_0 t).$$

Exercise 8.19 Verify this result. [*Hint*: You will need to note that in classical statistical mechanics, $\langle A\dot{A}\rangle = 0$.]

Performing the Fourier transform yields two delta functions—one at ω_0 and the other at $-\omega_0$. That is,

$$\text{abs}(\omega) \propto \delta(\omega - \omega_0), \qquad \omega > 0.$$

If more generally, the dynamics of A behaved as if A was one member in a system of coupled harmonic oscillators, abs (ω) would then be proportional to the density of states, or the concentration of normal modes in the system with frequency ω.

This example illustrates that absorption occurs when the monochromatic disturbance is tuned to the intrinsic frequencies in the system. There are many everyday examples of this phenomenon. Think of pushing a child on a swing or rocking a car that is stuck in a snow bank. In both cases, one attempts to match the intrinsic frequency (the pendulum motion of the swing in the first case, and the frequency of the suspension springs of the automobile in the second). When the match is achieved, you are able to efficiently impart energy to the system.

One final remark before leaving this section. All the formulas we have derived are correct for classical dynamical systems. The analysis of quantum systems is a bit more complicated, and the formulas differ somewhat from those shown here. However, the basic logical development and all the general qualitative conclusions we have drawn from the classical treatment remain true in the quantum mechanical case.

8.8 Friction and the Langevin Equation

One of the most familiar examples of non-equilibrium behavior and dissipation is the phenomenon of friction. In particular, a particle pushed with velocity v through a medium experiences frictional drag or force proportional to the velocity,

$$f_{drag} = -\gamma v.$$

Here, γ is the friction constant. The presence of a velocity-dependent force implies that a fluid will absorb energy as one drags a particle through the fluid. In other words, as a result of friction, energy is dissipated, and the medium will heat up. In this section, we describe a model for frictional dissipation.

The particular model corresponds to an oscillator or vibrating degree of freedom coupled to a bath. A schematic picture of the system is given in Fig. 8.12. The Hamiltonian we have in mind is

$$\mathcal{H} = \mathcal{H}_0(x) - xf + \mathcal{H}_b(y_1, \ldots, y_N),$$

where $\mathcal{H}_0$ and $\mathcal{H}_b$ are the oscillator and bath Hamiltonians, respectively, the oscillator variable x is coupled to the multitude of bath variables $y_1, \ldots, y_N$ through the force, f, which depends linearly upon the bath variables,

$$f = \sum_i c_i y_i,$$

where the c_i's are constants.

The oscillator and its coordinate, x, are the focus of our attention.

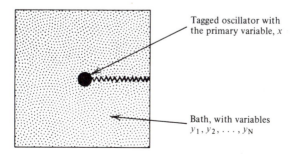

Tagged oscillator with the primary variable, x

Bath, with variables $y_1, y_2, \ldots, y_N$

Fig. 8.12. Oscillating particle coupled to a bath.

That is, it is the primary degree of freedom. The bath variables are secondary degrees of freedom. We will assume nothing in particular about $\mathcal{H}_0(x)$ other than to say that the oscillator is classical and if it was left to itself, it would conserve energy. That is,

$$\mathcal{H}_0 = m\dot{x}^2/2 + V(x),$$

hence, $d\mathcal{H}_0/dt = 0$ implies that $m\ddot{x} = -dV(x)/dx$.

We do, however, assume something very special about the bath. In particular, in the model we adopt, $\mathcal{H}_b$ is the Hamiltonian for a collection of harmonic oscillators. The most important results we derive employing this model may be more general than the simplified picture might suggest. We will not worry about the generality, however, because even without it, the model of a harmonic bath linearly coupled to a primary degree of freedom already captures many of the physical phenomena one encounters when considering dissipative systems. It is for this reason that the model is a very popular one in the physical sciences. It is often true, however, that great care must be taken to accurately approximate natural nonlinear systems with a harmonic bath.

The terminology "nonlinear" is used here to note that in general, the potential energy of a system gives rise to forces that are nonlinear functions of the coordinates. Harmonic systems, however, are linear. That is, the restoring force to any displacement of a harmonic system is a restoring force that is proportional to the displacement. As a consequence, stable harmonic systems exhibit only linear responses to disturbances of arbitrary size. In other words, if we consider the harmonic bath by itself, the theory of linear response analyzed in the previous sections becomes an exact theory and not simply one limited to very weak perturbations. From Sec. 8.6 we know that this response of the bath is characterized by correlation functions like

$$C_b(t) = \langle \delta f(0)\delta f(t) \rangle_b = \sum_{i,j} c_i c_j \langle \delta y_i(0)\delta y_j(t) \rangle_b,$$

where the subscript "b" is used to denote the properties of the pure bath (uncoupled to the oscillator). Since our choice of the coordinates $\{y_i\}$ is arbitrary, we can use those that are the normal modes. In that case, the $i \neq j$ terms in the last equality are zero, and $\langle \delta f(0)\delta f(t) \rangle_b$ can be expressed in terms of an integral involving the spectrum of normal mode frequencies for the bath. We will not need to exploit that representation in what follows. But we do, however, make use of the fact that a harmonic bath has no nonlinear response.

Given that fact, let us consider the time evolution of $f(t)$. Let $x(t)$ denote the time dependent x yet to be determined self-consistently once we know $f(t)$. This function $x(t)$ changes the evolution of $f(t)$

from what would be found in the pure bath, denoted by $f_b(t)$. The change is determined by the linear response formula,

$$f(t) = f_b(t) + \int_{-\infty}^{\infty} dt' \, \chi_b(t - t')x(t'),$$

where $\chi_b(t - t')$ is the response function for the pure bath,

$$\chi_b(t - t') = -\beta \, dC_b(t - t')/d(t - t'), \qquad t > t'$$
$$= 0, \qquad t < t'.$$

Here, we are simply reciting what was learned from the fluctuation-dissipation theorem in Secs. 8.5 and 8.6.

Now we turn our attention to the primary variable, x. According to Newton's law,

$$m\ddot{x} = \text{force}.$$

There are two contributors to that force. One comes from the oscillator potential, $V(x)$, and the other is from the bath. Combining both, we have

$$m\ddot{x}(t) = f_0[x(t)] + f_b(t) + \int_{-\infty}^{\infty} dt' \, \chi_b(t - t')x(t'),$$

where

$$f_0[x] = -dV/dx.$$

Notice that this equation of motion for $x(t)$ has a random or fluctuating force, $f_b(t)$, arising from the bath, and it has a term that is nonlocal in time. The latter arises because $x(t)$ influences the behavior of the bath, and this influence persists to later times. The extent of that influence is determined by the length of time over which correlations regress in the bath.

The nonlocal term can be reexpressed in terms of $C_b(t) = \langle \delta f(0)\delta f(t) \rangle_b$ by employing the fluctuation-dissipation theorem and integrating by parts (with care noting that the boundary terms are not zero). The result is

$$m\ddot{x}(t) = \bar{f}[x(t)] + \delta f(t) - \beta \int_0^t dt' \, C_b(t - t')\dot{x}(t')$$

where

$$\bar{f}[x] = -d\bar{V}/dx$$

with

$$\bar{V}(x) = V(x) - \beta C_b(0)x^2/2$$

and

$$\delta f(t) = f_b(t) - \beta C_b(t)x(0).$$

The statistics of the random fluctuating force $f_b(t)$ is Gaussian with mean value $\beta C_b(t)x(0)$ and variance $C_b(t - t')$. The function $\bar{V}(x)$ is the potential of mean force. That is, $\bar{f}[x]$ is the bath averaged force on the primary coordinate.

Exercise 8.20* Verify these formulas from the preceding development.

In deriving these formulas, one considers the case where the initial conditions correspond to the state of the system at $t = 0$. Since the coordinates and momenta of the bath are not characterized specifically, however, one averages over the $t = 0$ state of the bath. Only the primary variable $x(t)$ remains. After a transient period, the bath at time t is uncorrelated with the initial state, and $f_b(t)$ becomes $\delta f(t)$, a Gaussian fluctuating force of zero mean.

This equation of motion for $x(t)$ is the *generalized Langevin equation* derived by Robert Zwanzig. Notice that the term nonlocal in time gives a velocity-dependent force, and the size of this force is determined by the extent of the fluctuating forces associated with the coupling to the bath. Thus, we see clearly that friction is a manifestation of fluctuating forces. This connection between frictional dissipative forces and the fluctuating force autocorrelation function is sometimes called the *second fluctuation-dissipation theorem*, a terminology introduced by Ryogo Kubo.

Historically, the Langevin equation was used to analyze the random motion of a tagged macroscopic particle moving through a thermally equilibrated molecular medium. This is the phenomenon known as *Brownian motion* (named after Robert Brown, who observed the perpetual irregular motion of pollen grains immersed in a fluid). In that case, there is no average force, so that $\bar{f} = 0$. Furthermore, the relaxation time for fluctuating forces in the bath is negligibly short compared to the time and distance scales over which one observes so-called Brownian particles. That is,

$$\int_0^t dt' \, C_b(t')\dot{x}(t - t') \approx \dot{x}(t) \int_0^\infty dt' \, C_b(t').$$

This approximation removes the memory effects (nonlocality in time) from the equation of motion. It is often called a *Markovian* approximation. Under these circumstances of locality in time and $f_0 = 0$, the equation of motion for a coordinate of the tagged particle becomes the traditional Langevin equation

$$m\ddot{x}(t) \approx f_b(t) - \gamma v(t),$$

where $v(t) = \dot{x}(t)$ and

$$\gamma = \beta \int_0^\infty dt \langle \delta f(t) \delta f(0) \rangle_b.$$

From this simplified equation, the following picture of particle dynamics is drawn: The particle experiences random forces that buffet the particle about. These forces arising from the bath add energy to the particle, but excessive increases in kinetic energy are removed by frictional dissipation (the velocity-dependent force). To further expand on this interpretation, we can compute the velocity correlation function for the tagged particle. This is done by multiplying the Langevin equation by $v(0)$ and averaging. Note that since $f_b(t)$ is independent of the primary variable,

$$\langle v(0) f_b(t) \rangle = \langle v \rangle \langle f_b \rangle = 0.$$

Hence

$$m \langle v(0) \dot{v}(t) \rangle = -\gamma \langle v(0) v(t) \rangle$$

or

$$\frac{d}{dt} \langle v(0) v(t) \rangle = -(\gamma/m) \langle v(0) v(t) \rangle.$$

The solution to this differential equation is

$$\langle v(0) v(t) \rangle = \langle v^2 \rangle e^{-(\gamma/m)t}.$$

Therefore, the exponential relaxation of the velocity correlations considered in Exercise 8.13 arises from the Langevin equation in the Markovian approximation.

The general form of the Langevin equation plays an important role in computational physics and chemistry where it is used as the theoretical basis for performing *stochastic molecular dynamics* simulations. You can use these words and our discussion of the Langevin equation as clues to guess what is referred to by such simulations. But for this book, I have told you everything I planned to. Hopefully, what has been explicitly written here or suggested in the Exercises provides the foundation and motivation to rapidly pursue advanced writings on stochastic dynamics as well as many other areas of this rich and ever-changing field, statistical mechanics.

Exercise 8.21* Here we use the Langevin equation to discuss the vibrational spectra of a harmonic oscillator in a condensed phase. That is, assume $V(x) = \frac{1}{2} m \omega_0^2 x^2$, and imagine that x is proportional to a dipole that can couple to an electric field and therefore absorb energy. Show that the Langevin equation gives

$$\frac{d^2}{dt^2} \langle x(0)x(t) \rangle = -\bar{\omega}^2 \langle x(0)x(t) \rangle$$

$$-(\beta/m) \int_0^t dt' \, C_b(t-t') \frac{d}{dt'} \langle x(0)x(t') \rangle,$$

where $\bar{\omega}$ is the average frequency of the oscillator. It is shifted from ω_0 by the bath according to the formula

$$\bar{\omega}^2 = \omega_0^2 - (\beta/m)C_b(0).$$

Linear differential-integral equations are compactly solved with the method of Laplace transforms. Let $\tilde{C}(s)$ denote the Laplace transform of $\langle x(0)x(t) \rangle$. Show that

$$\tilde{C}(s) = \frac{s + (\beta/m)\tilde{C}_b(s)}{s^2 + \bar{\omega}^2 + s(\beta/m)\tilde{C}_b(s)} \langle x^2 \rangle$$

where $\tilde{C}_b(s)$ is the Laplace transform of $C_b(t)$. Assume that the bath fluctuation correlations are exponential—that is,

$$C_b(t) = C_b(0)e^{-(t/\tau)},$$

and therefore

$$\tilde{C}_b(s) = \frac{C_b(0)}{s + \tau^{-1}}.$$

With this formula compute the cosine Fourier transform of $\langle x(0)x(t) \rangle$ by showing that

$$\int_0^\infty dt \, \cos(\omega t) \langle x(0)x(t) \rangle = \text{Re } \tilde{C}(i\omega).$$

Use this result to describe the absorption spectra of a tagged oscillator in terms of the parameters ω_0, τ^{-1} and $C_b(0)$. Comment on how the spectra line shape depends upon the spectrum of modes in the bath.

Additional Exercises

8.22. Draw labeled pictures and discuss the qualitative features of the following autocorrelation functions:

(a) $\langle \mathbf{v}(0) \cdot \mathbf{v}(t) \rangle$ where $\mathbf{v}(t)$ is the velocity of an argon atom in the gas phase;

  (b) $\langle \mathbf{v}(0) \cdot \mathbf{v}(t) \rangle$ for an argon atom in the solid phase;

  (c) $\langle v^2(0)v^2(t) \rangle$ for an argon atom in the solid phase;

  (d) $\langle \mathbf{u}(0) \cdot \mathbf{u}(t) \rangle$ where $\mathbf{u}(t)$ is the unit vector in the direction of the principal axis of a tagged CO molecule in a solid lattice.

Discuss the sizes of the different time scales for each of these functions.

8.23. Consider a dilute solution of ions. An applied electric field creates a net flow of charge. After the field is turned off, the flow dissipates. Estimate the time it takes for this relaxation to equilibrium to occur in terms of the self-diffusion constant and mass of the ions.

8.24. According to Stokes' law (which pictures a fluid as a viscous continuum), the self-diffusion constant of a spherical particle of diameter σ is

$$D = (2 \text{ or } 3)^{-1}(1/\pi\beta\sigma\eta),$$

where η is the shear viscosity, and the factors of 2 or 3 depend on whether the solvent slips past or sticks to the diffusing solute, respectively. According to this formula, the friction constant in the Langevin equation is proportional to η. Use this equation to estimate the typical size of D for a molecule of 5 Å diameter dissolved in a liquid with $\eta \approx 10^{-2}$ poise (a standard value for most ordinary liquids and water, too).

8.25. Consider particles in a liquid for which D is $10^{-5} \text{ cm}^2/\text{sec}$. Determine the percentage of particles which in 5 psec have moved more than 5 Å from their initial positions.

8.26. Consider a reaction coordinate q that moves in a one-dimensional bistable potential $V(q)$. Divide phase space for this degree of freedom into three regions as illustrated in Fig. 8.13. The non-equilibrium concentrations of species in the

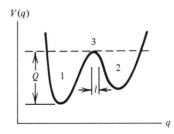

Fig. 8.13. Bistable potential and three regions.

three regions are $c_1(t)$, $c_2(t)$, and $c_3(t)$. A reasonable rate law description of the dynamics is

$$\frac{dc_1}{dt} = -k_{31}c_1(t) + k_{13}c_3(t),$$

$$\frac{dc_2}{dt} = -k_{32}c_2(t) + k_{23}c_3(t),$$

and

$$c = \text{constant} = c_1(t) + c_2(t) + c_3(t).$$

Here, k_{ij} is the rate for going from region j to region i. From detailed balance we have

$$k_{13} \quad \text{and} \quad k_{23} \gg k_{31} \quad \text{and} \quad k_{32}$$

since

$$\left(\frac{k_{31}}{k_{13}}\right) = \frac{\langle c_3 \rangle}{\langle c_1 \rangle} \approx e^{-\beta Q}$$

and we assume the barrier height, Q, is large compared to $k_B T$.

(a) Use the simple kinetic model to compute the time-dependence for

$$\Delta c_1(t) = c_1(t) - \langle c_1 \rangle.$$

(b) Show that if $\exp(-\beta Q) \ll 1$, the relaxation is dominated by single relaxation time, τ_{rxn}, and that

$$\tau_{rxn}^{-1} \approx k_{31} \quad \text{or} \quad k_{32}.$$

(c) Discuss the transient behavior of $\Delta c(t)$ or its time derivative and show how this behavior disappears in a time of the order of k_{13}^{-1} or k_{23}^{-1}.

(d) Discuss the behavior of this model in the context of the reactive flux $k_{BA}(t)$ described in the text.

(e) Suppose the average time it takes to go from region 3 to region 1 or 2 is determined by the time it takes the reaction coordinate to diffuse the distance l. Justify

$$\bar{k}_{21} \propto (1/\eta)e^{-\beta Q},$$

where η is the viscosity coefficient or friction constant felt by the reaction coordinate, and $\bar{k}_{21}$ is the effective rate constant (i.e., the observed rate constant) for going from region 1 to region 2.

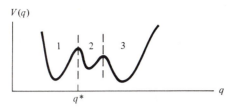

Fig. 8.14. Tristable potential.

8.27. Consider a tristable potential (shown in Fig. 8.14) for a reaction coordinate q. Derive a microscopic expression for the rate constant k_{21}, which is defined by the phenomenological rate equations

$$dc_1/dt = -k_{21}c_1(t) + k_{12}c_2(t),$$
$$dc_2/dt = k_{21}c_1(t) - k_{12}c_2(t) + k_{23}c_3(t) - k_{32}c_2(t),$$

and

$$c_3(t) = c - c_1(t) - c_2(t).$$

8.28. Consider an aqueous solution with argon dissolved at low concentration. At room temperature (300°K), the viscosity of water is $\eta = 0.01$ poise, and the self-diffusion constant for the argon atoms is 1×10^{-5} cm^2/sec. The mass of an argon atom is 40 amu, and the mass of a water molecule is 18 amu.

(a) What is the root-mean-square velocity of an argon atom in the vapor that is at equilibrium with the solution?

(b) What is the root-mean-square velocity of an argon atom in the solution?

(c) Estimate the time it will take an argon atom to move 10 Å from its position at time zero in the vapor.

(d) Estimate the time it will take an argon atom to move 10 Å from its initial position in the solution.

(e) Estimate the time it will take for an initially prepared non-equilibrium velocity distribution for the argon atoms in solution to relax to the equilibrium distribution.

(f) By employing high pressure equipment, the water is brought to a thermodynamic state at $T = 300°K$ for which $\eta = 0.02$ poise. For this state, what are the answers to the questions posed in parts (b) and (d)? [*Hint*: See Exercise 8.24.]

Bibliography

Forster's informally written monograph presents a complete survey of linear response theory, including the formal properties of correlation functions, the connection between measurements and correlation functions, and the Langevin equation, too:

> D. Forster, *Hydrodynamic Fluctuations, Broken Symmetry and Correlation Functions* (Benjamin Cummings, Reading, Mass., 1975).

McQuarrie and Friedman's texts, written from a chemistry perspective, contain chapters discussing time correlation functions:

> D. McQuarrie, *Statistical Mechanics* (Harper & Row, New York, 1976).

> H. L. Friedman, *A Course in Statistical Mechanics* (Prentice-Hall, Englewood Cliffs, N. J., 1985).

Ryogo Kubo is one of the pioneers of linear response theory especially its implications to the theory of molecular relaxation (a subject barely touched upon here and only in our Exercise 8.21). The book created by Kubo and his co-workers has much to say on this topic:

> R. Kubo, M. Toda, and N. Hashitsume, *Statistical Physics II*: *Nonequilibrium Statistical Mechanics* (Springer-Verlag, New York, 1985).

A general discussion of the connection between correlation functions and relaxation equations such as the diffusion equation of Sec. 8.4 is given in Chapter 11 of

> B. J. Berne and R. Pecorra, *Dynamic Light Scattering* (John Wiley, New York, 1976).

Index